Imagining Solar Energy

Explorations in Science and Literature

Series Editors

John Holmes, Anton Kirchhofer and Janine Rogers

Explorations in Science and Literature considers the significance of literature from within a scientific worldview and brings the insights of literary study to bear on current science. Ranging across scientific disciplines, literary concepts, and different times and cultures, volumes in this series will show how literature and science, including medicine and technology, are intricately connected, and how they are indispensable to one another in building up our understanding of ourselves and of the world around us.

Forthcoming titles

Biofictions, Josie Gill
The Diseased Brain and the Failing Mind, Martina Zimmermann
Narrative in the Age of the Genome, Lara Choksey
Fictions of Prevention, Benedetta Liorsi
The Social Dinosaur, Will Tattersdill

Imagining Solar Energy

*The Power of the Sun in Literature,
Science and Culture*

Gregory Lynall

BLOOMSBURY ACADEMIC
LONDON • NEW YORK • OXFORD • NEW DELHI • SYDNEY

BLOOMSBURY ACADEMIC
Bloomsbury Publishing Plc
50 Bedford Square, London, WC1B 3DP, UK
1385 Broadway, New York, NY 10018, USA

BLOOMSBURY, BLOOMSBURY ACADEMIC and the Diana logo are trademarks
of Bloomsbury Publishing Plc

First published in Great Britain 2020

Cover design: Toby Way
Cover image © Getty Images

A catalogue record for this book is available from the British Library.

A catalog record for this book is available from the Library of Congress.

ISBN: HB: 978-1-3500-1097-0
ePDF: 978-1-3500-1099-4
eBook: 978-1-3500-1098-7

Series: Explorations in Science and Literature

Typeset by Deanta Global Publishing Services, Chennai, India

To find out more about our authors and books visit www.bloomsbury.com
and sign up for our newsletters.

Contents

Illustrations

Acknowledgements

This book would not have been completed without the generous support of the Leverhulme Trust and the School of the Arts, University of Liverpool. My colleagues in the Department of English at Liverpool have played crucial roles. Paul Baines, Matthew Bradley, Alex Broadhead, Andrew Duxfield, Jonathan Roberts, David Seed and Sam Solnick have all read portions of the book, and made invaluable comments. Many others – including Nandini Das, Siobhan Chapman, Michael Davies, Alex Harris, David Hering, Chris Pak, Chris Routledge, Jill Rudd and Andy Sawyer – have offered advice and enthusiasm at decisive moments. The Literature & Science Hub at Liverpool was very fortunate to host Kelly Sultzbach as a Fulbright scholar in spring 2019, and her comments were essential in honing the introductory and concluding chapters. Alice Monter and Simon Logan patiently corrected my French and German, respectively. Marcus Walsh and Wendy Perkins are always there whenever I need them. I discussed the project early on with the late Nick Davis, and wish that he were still around to make it better. Early modern aspects of the book have been presented at various fora involving Liverpool's Eighteenth-Century Worlds Research Centre – always a stimulating and fun group to be part of, although it will never be quite the same without the incredible Kate Marsh. Georgina Endfield, Jon Major, Chris Pearson and Neil Winterton have all helped me to extend my knowledge. Anna Burton and Harriet Barton each spent some time as research assistants on this project, and their intelligence and flair was much appreciated; Anna also read late drafts of several chapters, and corrected several errors.

This book would also not exist without the British Society for Literature and Science. I've had conversations with many generous and wise members over the years, but special appreciation should go to Alex Campbell, Rachel Crossland, Folkert Degenring, Barri Gold, Jerome de Groot, Alice Jenkins, Peter Middleton, Jim Mussell, Matt Paskin, Sharon Ruston, Sally Shuttleworth and Martin Willis for their advice and suggestions. Jim Scown kindly read a chapter for me, and made enlightening comments. Further afield, Cedric Carles, Frank James, Simon Schaffer, Larry Stewart, Sophie Vasset and Tania Woloshyn have all stimulated ideas or provided important information.

Thierry Lalande was an exceedingly generous host during my visit to the Paris Musée des arts et métiers, and it was amazing to see for real Buffon's burning-mirrors and one of the Mouchot-Pifre solar generators. For bringing this book into being, sincere thanks should go especially to the series editors Janine Rodgers, John Holmes and Anton Kirchhofer, whose general oversight of the whole project has been invaluable.

Papers towards this book have been delivered at the annual conferences of the British Society for Literature and Science, the British Society for Eighteenth-Century Studies, ASLE-UKI, SLSA-EU, SRUK-CERU and in research seminars at the universities of Liverpool, Northumbria, Paris-Diderot, Liverpool Hope and Keele. The 'Transforming Objects' conference at Northumbria University, in 2012, was fundamental to the book's initial conception, and I thank the organizers Nicole Bush and Anna Hope. Other events that have shaped the book include the 'Theories and Uses of Light in British Arts' conference at Paris-Diderot in 2014, organized by Sarah Gould and Diane Leblond; the EPSRC 'Fundamentals of New and Sustainable Photovoltaics' 2015 doctoral training course, organized by Ken Durose and Rob Treharne; the 'Power at the Edge' workshop at Liverpool in 2018, organized by Jonathan Hogg and Marianna Dudley; and the 'Narrative Science' workshop at LSE in 2019, organized by Dominic Berry. I've also had the privilege of delivering several public talks related to this work, at the National History Museum, LightFest at the Library of Birmingham (organized by Aston Institute for Photonic Technologies), the Wellcome Collection, METAL Liverpool, Tate Liverpool, and Holywell local history society. It has also been a pleasure to disseminate my work via public exhibitions and creative workshops, at Tate Liverpool and Alder Hey Children's Hospital, and I am extremely grateful to all of those who have helped me along the way, including Melissa Raines, Philippa Holloway, Bernadette McBride and Natalie Hanna. Enlightenment visual culture associated with solar technology is especially compelling (and in some cases, riotous), and it was thrilling to display this particular material within these public exhibitions (and within this book).

Small portions of the book have been published previously as '"Bundling Up the Sun-Beams": Burning Mirrors in Eighteenth-Century Knowledge and Culture', in a special issue of *Journal for Eighteenth-Century Studies*, 36: 4 (Winter 2013), 477–90, and 'The Freedom of the Solar Cell: Energies of the Sun across the Long Twentieth Century', in *Energy in Literature: Essays on Energy and Its Social and Environmental Implications in Twentieth and Twenty-First Century Literary Texts*, ed. Paula A. Farca (2015), pp. 243–57. My thanks to the editors, Sarah Easterby-Smith and Emily Senior, and Paula Farca, respectively, for encouraging my work at an early stage, and to the presses for allowing republication of this material.

I am grateful to Sara Norja for allowing me to quote from 'Sunharvest Triptych', in *Sunvault: Stories of Solarpunk and Eco-Speculation*, eds. Phoebe Wagner and Brontë Christopher Wieland (Upper Rubber Boot, 2017). Extracts from Derek Mahon, 'Its Radiant Energies' from *Life on Earth* (Gallery Books, 2008) reproduced by kind permission of the author and The Gallery Press. Quotations from *Blinded by the Sun* © Stephen Poliakoff, 1996, *Blinded by the Sun & Sweet Panic*, Methuen Drama, an imprint of Bloomsbury Publishing Plc. Extracts from *Lord of the Flies*, Copyright the Estate of William Golding. Reproduced by permission of Faber & Faber Ltd/Penguin Random House. Extracts from 'Solar', Copyright the Estate of Philip Larkin, first appeared in *High Windows* (London: Faber & Faber, 1964). Reproduced by permission of Faber & Faber Ltd/Farrar, Straus and Giroux. Extracts from 'Sunlight', Copyright the Estate of Thom Gunn, first appeared in *Moly* (London: Faber & Faber, 1971). Reproduced by permission of Faber & Faber Ltd/Farrar, Straus and Giroux. Extracts from *The Burning Glass* © Charles Morgan, 1954, by kind permission of Oberon Books Ltd. Third-party copyrighted materials are displayed in the pages of this book on the basis of fair dealing for the purposes of criticism and review and aim to be in accordance with international copyright laws, and are not intended to infringe upon the ownership rights of the original owners.

University of Liverpool Special Collection & Archives and particularly the Science Fiction Foundation Collection have been crucial to my research, and particular thanks should go to Robyn Orr, Katy Hooper and Nicola Kerr. Visits to the British Library and the libraries of the Royal Astronomical Society, Royal Institution, Royal Society, the University of Birmingham and the Wellcome Collection were also essential. I acknowledge the kind permission of the Royal Astronomical Society and Royal Institution in allowing me to quote from their manuscript holdings. Many thanks to Dan Phillips for expertly photographing the images in my possession.

Finally, thanks must go to my wonderful family, for accompanying me on various trips as I've hunted for books or given talks, and for allowing me to spend much of my 'free time' behind a desk. Hopefully we'll find more hours to play in the sunshine now.

Series preface

In spite of the myth of the 'Two Cultures', science and literature have always been shaped by one another. Many of our most powerful scientific concepts, from natural selection to artificial intelligence, from germ theory to chaos theory, have been formed through the careful – and sometimes careless – use of written language. Poets, novelists, playwrights and journalists have taken up scientific ideas, medical research and new technologies, exploring them, reworking them, at times distorting or misjudging them, but always shaping profoundly the wider culture's understanding of what they mean. This intimate and productive relationship between literature and science generated a steady stream of insightful scholarship and commentary throughout the twentieth century and has grown into a substantial field of study in its own right since the turn of the millennium. Where the idea of 'Two Cultures' does still have a hold, however, is in academic disciplines themselves. In schools and universities, we study science and arts subjects in different classrooms, taught by different people with different expectations. Literature and science studies has, so far, been largely a sub-discipline of literature, with only rare contributions from or addressed to scientific experts. In a world of ever-increasing specialization, failure to communicate across these disciplinary divides risks failing to appreciate the contribution that the study of literature can make to our understanding of science, medicine and technology, the uses that science makes of images, narratives and fictions, and the insights that scientists can bring to bear on literature and on culture at large.

Explorations in Science and Literature aims to speak across this divide. It has a particular mandate to bring the insights of literary study to bear on science itself; to consider the significance of literature from a scientific point of view; and to explore the role of literature within the history of science. The books, therefore, examine the complex interrelations between science and literature in cross-disciplinary ways. They are written equally for scholars and students of literature and for scientists and science students, but also for historians and sociologists of science, as well as general readers interested in science and its place in culture and society. By showing how each field can be enhanced by a knowledge of the others, we hope to enrich scientific as well as literary research and to cultivate a new cross-disciplinary approach to fundamental questions in both fields.

The series will encompass topics from across the physical, biological and social sciences, and medicine and technology, wherever literature can inform our understanding of the science, its origins and its implications. It will also include books on literary forms and techniques that are informed by science, as well as studies that consider how science itself has been articulated. Along with literature in the broad sense of written texts, books in the series will also consider other cultural forms including drama, film, television and other arts and media.

John Holmes, Anton Kirchhofer and Janine Rogers

Introduction:
Bringing the Sun into focus

In 1716 appeared *To the Right Honourable the Mayor and Aldermen of the City of London: The Humble Petition of the Colliers, Cooks, Cook-Maids, Blacksmiths, Jack-makers, Brasiers, and Others*. This pamphlet was apparently written on behalf of tradespeople disgruntled by a group of '*Virtuosi*' seeking to procure the monopoly of an optical device. Calling themselves 'CATOPTRICAL VICTUALLERS', these suppliers of reflected light 'presumed, by Gathering, Breaking, Folding, and Bundling up the Sun-Beams, by the help of certain *Glasses*, to Make, Produce, and Kindle up several New *Focus*'s or Fires ... and thereby ... to perform all the Offices of Culinary Fires'. This would 'monopolize the Beams of the Sun', and the capital- and labour-saving features of the technology would destroy the London economy, reducing cooks and blacksmiths 'to Beggary' and forcing colliers out of business, just at the moment coal usage was beginning to radically transform Britain. This petition, of course, is a parody, and was reprinted in the *Miscellanies* (1732) published by satirists Alexander Pope (1688–1744) and Jonathan Swift (1667–1745).[1] It is likely, however, to have been written by their friend, Dr John Arbuthnot (1667–1735), the mathematician, physician and fellow of the Royal Society.[2] Despite its 300-year vintage, *The Humble Petition of the Colliers*, as it has become known, emphasizes questions of energy usage and competition that seem incredibly modern. It appears to be the first known case, in fact or fiction, of a fossil fuel supplier lobbying a government against the use of renewables. Imagining the fantasies and fears that surround moments of energy transition, the pamphlet identifies the winners and losers of the resulting economic and social changes. It also reminds us of the public suspicions and expectations regarding the agency of science within society. In reality, the colliers would not have to worry about solar technologies for quite some time – the UK's National Grid eventually went a whole day without coal in 2017.[3]

The future of energy poses challenges not only scientifically and technologically but also politically and culturally.[4] Social processes, along with political and cultural meanings of energy, are playing their part in determining how humanity will transition away from polluting and depleting carbon-intensive resources. A profound dependence upon fossil fuels – involving a complex network of extraction, distribution and consumption – has constructed industrial modernity and the globalized economy.[5] There may need to be a transition in values and practices as well as technologies, with our societies re-conceptualized within alternative energy paradigms and systems that challenge accepted notions of space, community, lifestyle and human interaction with the environment – and involving transformation within a rapid timeframe. Fortunately, as *The Humble Petition* shows us, literature and other forms of art and culture have been involved in this visionary process for many years, and the scenarios of energy production and consumption they have generated – operating from multiple perspectives and scales, from the individual psyche to the planetary – have the potential to inform sociopolitical, scientific and technological choices.

Scholars in the emerging field of the energy humanities are uncovering the ways in which hydrocarbons have structured modern life and thought.[6] Although the world's transition to renewables is only just beginning, it is also important to recover the history and culture of these supposedly 'new' and 'alternative' forms of energy, and reflect upon them within the light of contemporary concerns and approaches. Moreover, this is not the first time communities have needed to transition to different forms of energy in order to deal with societal difficulties, and demands upon current resources. As President Carter stated in his famous national address on energy in 1977, 'we must look back into history to solve our energy problems', to study how previous challenges were met intellectually, politically and culturally.[7] The cosmic irony, of course, is that in the twenty-first-century shift to renewable energy we are ultimately returning to sources that powered much of the world up until two hundred years ago. However, the dilemmas encountered now are on very different scales to those faced previously, encompassing the survival of habitats and species across the entire planet, threatened by the results of past and present energy choices. Approaching peak oil is not the only problem. The results of previous human activities can now be witnessed geologically and assigned scientific nomenclature (the Anthropocene), and the alarming environmental legacy of previous and current energy consumption, particularly the level of greenhouse gases in the atmosphere, can be detected and forecasted.[8] Science and technology are deeply implicated in the causes of anthropogenic climate change, but have identified this crisis, and will spearhead the solution. Politics and economics are, of course, also crucial, and the moment renewable energy achieves grid parity around the globe will mark a watershed. Yet to adequately appreciate and, most importantly, act upon the predicted consequences of our energy-related behaviours involves the work of the collective imagination, and literature and art may assist these creative leaps.

This book considers the roles literature and culture have played in imagining, mythologizing and reflecting the possibilities of harnessing solar energy. It is unique in its literary and artistic focus, and establishes the importance of the concept of solar energy to many literary works for the first time. Locating the history of endeavours to capture solar radiance as a focal point between science and the imagination, this study argues that the literary, artistic and mythical resonances of solar power have not only been inspired by but also cultivated and sustained its scientific and technological development. It explores the many tangible connections between the literary and the scientific/ technological, and thinks through the ways in which scientific discourses have narrativized knowledge about solar power. But the book's argument is unashamedly structured around the questions and contexts raised by works of the imagination, which have represented, figuratively transformed and formally appropriated solar energy in order to illuminate and ignite volatile scientific, cultural and political matters. Recent work in the energy humanities has drawn attention particularly to the influence of energy systems upon culture and politics.[9] This book instead sees a more reciprocal relationship between solar energy and culture, identifying many moments when fictionalizations of solar power have interposed within the history of science and technology, inspiring research or at least framing the meaning of knowledge production. Indeed, in some cases, literary and scientific endeavours are indistinguishable from one another.

While acknowledging that the future of energy is a global issue, the book explores, in particular, the relationship between British literature and science, and the cultural exchange between Britain, continental Europe and the United States. Given the relatively poor insolation of the British Isles, it might seem odd to discover such a wide variety of solar imaginaries emanating from this location, but the propensity to inclement weather has perhaps encouraged writers, artists, scientists and engineers alike to invent ways to utilize and celebrate the bringer of light. The dominance of British science over much of the eighteenth and nineteenth centuries is, of course, another factor. Yet the American science writer Daniel Behrman also found 1970s UK to be a place of particular solar cultural investment: 'If the Sun's bounty were meted out on the basis of the interest, even fascination, that it awakens here below, Great Britain would surely enjoy a surfeit.'[10] Despite the Anglo-centric focus, it is hoped that this work will stimulate research that uncovers solar imaginaries within other literatures and cultures. The book's temporal span, meanwhile, is wide, covering the Renaissance up to the present day. Over this time, there has been an ebb and flow in scientific and literary interest in solar energy, and in the kinds of genre, mode and form used to represent and imagine solar technologies. Nevertheless, patterns and continuities do emerge when looking longitudinally, although they are far from constituting a Whiggish history.

The notion of 'solar energy' is historically specific in itself, a product of the nineteenth century, and it might appear anachronistic to trace its cultural and scientific presence prior to that time. Yet the Sun's emissions have been described using a lengthy number of 'solarnyms', including sunlight, sunbeams, sun-fire, lumen, astral virtues, Promethean fire, rays, radiance, sun-like lightenings, photons, and 'electromagnetic disturbance'.[11] Many of them predate, but survive alongside, 'solar energy' and continue to contribute to its meaning. The adjective and noun 'renewable', meanwhile, has pertained to natural, sustainable forms of energy from around the late 1940s, but came into common parlance in the 1980s.[12] All of these solarnyms attempt to grasp at solar energy's status as a 'hyperobject' of an almost infinite spatial and temporal scale and multiple interpretative frame, and beyond rationalization in its totality.[13] Nevertheless, the flow of solar energy has been shaped into one of our most fascinating stories. The temperature at the Sun's core is 15 million degrees, and it takes sunlight 170,000 years to travel from there to the surface. Photons reach Earth, 150 million kilometres away, in 8 minutes, with every square metre of our planet bombarded by 100 billion photons every second.[14] Almost all processes upon Earth are ultimately powered by the Sun (directly or indirectly), making it the primary rather than an 'alternative' source of energy. Yet the Sun has a unique power not only physically but also symbolically, and its uncanny energies (both everyday and ultimately alien) reflect upon, and are refracted through, human material and textual productions.

Over the last forty years or so, the concept of solar energy has come into stark focus in two ways: the unwelcome increased greenhouse effect, caused by the accelerated gaseous absorption of infrared radiation, and its valued harnessing as an 'alternative' or 'renewable' source of power. However, the artificial capture of solar energy has a much longer history, and comprises many different forms of technology. This book principally engages with 'active' solar technologies – those which seek to intensify or transform the Sun's energy, such as photovoltaic cells and concentrating mirrors – rather than 'passive' forms, such as water-heating systems or large south-facing windows. These active forms, as will be explored, have stimulated the cultural imaginary in profound and various ways. The literary and artistic works discussed are also predominantly 'active' or explicit in approaching solar energy as a theme, interspersed with some less 'conscious' encounters.[15]

Each energy system has its own 'conceptual and experiential framework' to some extent: comprising connotations, metaphors and so forth, which are unique to each form.[16] However, this book maintains awareness of the relational meanings between different types of technology and energy, and of the potency of solar power as a figurative substitute for other sources (see especially Chapter 6). Solar devices did not develop in isolation, and their stories are entwined both practically and imaginatively with, for instance, optics, astronomy, telegraphy, photography and phantasmagoria. Some works, particularly those in the satiric mode, exploit the differences in scale between kinds of solar technology (see Chapter 2). Meanwhile, solar's position as an 'alternative' energy source, in dialogue with its culturally dominant rivals (coal, oil and nuclear energy), frames many of the ways in which the Sun's power is imagined and appropriated. A case in point is British Petroleum's rebrand as 'Beyond Petroleum' in 2000. Replacing its shield logo with a white, yellow and green Sun, the company attempted to publicize its green credentials, moving from stately, strong and protecting to open, abundant and ecological. Greenpeace were quick to accuse BP of greenwashing, arguing that they had spent more on this public relations exercise than on renewables over the whole of the previous year.[17] Fossil fuels are, of course, ultimately solar in origin – a fact that has conditioned attitudes to hydrocarbons from the nineteenth century onwards (see especially Chapter 4) – but BP's tendentious solar aesthetics were almost beyond parody.

The scale of the solar energy technologies that the book explores ranges from the immense solar power space station down to the handheld burning-glass. Despite their ostensible differences, each of the technologies operates as a form of the 'technological sublime' – a multifarious term, but one which generally refers to the ability of artificial objects to create experiences of awe and wonder within their spectators. Sublime encounters with solar technologies often go beyond the aesthetic moment, seeming to produce profound and sustained psychological effects. It is argued that the technological sublime manifests the power of the intellect in subjugating matter.[18] Solar technologies are, therefore, perhaps the ultimate symbols of the triumph of enlightened knowledge. They not only suggest the controlled extension of the most powerful and quintessentially sublime object in our cosmic proximity but also forcefully evoke in material form the traditional, Platonic association of the Sun with the light of reason.[19] Yet these dazzling devices are also mediators of a transcendental power beyond rationalization, whose ethereal presence breaks the bonds of the solar cell. It is when these technologies are misused or out of human control, when they have the potential to annihilate, that they are at their most sublime. Curiously, the handheld convex lens sometimes finds itself celebrated as the most exquisitely sublime solar technology, because it suggests access to unearthly, pure and supposedly limitless power that we ourselves can touch. This 'micro-sublime' object puts the Sun's power into the palm of the hand, perhaps the ultimate anthropocentric fantasy, liberatory and tyrannical in equal measure. Yet all of these solar technologies pale in comparison to the uncanny body of light they mediate, and the symbolic energy of this stellar object itself resonates across the spectrum of the works considered here. Our understandings of nature, energy, the environment and ourselves are bound up with the technologies we use. Adopting divine figurations from biblical, Classical and other traditions, burning-glasses have functioned as metaphors for Enlightenment, rationality, revelation, restoration and renewal. Investing human reason with an elemental force, solar technologies have offered conceptual frameworks through which to consider the place of natural knowledge in culture. Yet an object does not project a single discourse,[20] and technologies of the Sun are no exception in embodying particular, but multiple meanings. They can illuminate and dazzle, create and destroy, confer reason and delusion, and produce knowledge and wonder. They remind us of not only the Sun's creativity but also its destructive potential; of the Earth's eventual combustion under an expanded Sun; and of the accelerated greenhouse effect which may rob the planet of its life forms many years prior to that total destruction.

Solar power's meanings have also varied across time. We cannot essentialize contemporary solar devices or narratives, and then look back in history to find a single solar energy culture. The technologies and stories of solar power, therefore, need to be historicized appropriately within their specific contexts. Yet it will become apparent during the course of this book that first, many solar energy narratives often seek to emphasize mythopoeically the primal, ahistorical, universal nature of Sun-harnessing devices, and second, the concepts of earlier times still have a bearing upon how we imagine and understand solar energy today, influencing how humanity is responding to the climate crisis. Many works of 'sunlit', right up until the twenty-first century, self-consciously contemplate Promethean (and sometimes Archimedean) scenarios, demonstrating the intricate historical and cultural shaping of our understandings of energy systems. These discourses have structured not only humanity's anthropocentric relationship with the Sun and the environment as a whole but also our sense of what 'science' is, and who 'performs' it. In these cases, science is framed as humanity's rightful uncovering of nature's secrets, in momentary acts of discovery achieved by enlightened, individual *men* of scientific 'genius', granting our species access to immense forms of power. Solar energy narratives, therefore, are often characterized by discursive strategies that seek to deny the social circulations of knowledge in which the technologies are entangled, and to inflate the agency of their achievements. Not all stories of solar power operate within the heroic mode, however. Satirically and tragically, they sometimes focus on deluded characters, normally scientists, whose individualism and all-pervasive attention to the power of reason (often represented as a solar obsession) is often more a debilitating than liberating attribute. The Promethean narrative is subverted in other ways, too. Fictional solar energy scientists are predominantly male, conforming to the typical gendering of scientific labour historically, particularly in the physical sciences. Yet in the eighteenth-century stories, the most powerful wielder of solar power is the goddess Astraea, whose actions are apocalyptic in both revelatory and revolutionary ways. In the twenty-first century, new artistic modes (particularly 'Solarpunk') are attempting to challenge the conventional, patriarchal solar-sexual politics.

One of the most striking things to have come to light from examining solar power historically is its flexible technopolitics.[21] Solar energy technologies do not embody particular identities or politics inherently, but have been adopted across the political and cultural spectrum, although during certain periods they have been more likely to be associated with a specific ideology than others. The absolutist Sun kings of the seventeenth century, including Louis XIV himself, basked in the glory of burning-mirrors as part of their divine right propaganda, while renewable technologies are now more likely to find succour within left-wing political movements. Solar power is also an issue of global citizenship. Given the international variation in solar influx, the implementation of solar energy technologies will prove to be most useful in equatorial, often developing countries that have only benefited minimally from fossil fuels. 'Nature' has intervened, in this case, in ensuring the dispensation of energy justice between nations, and yet this relies upon fair distribution of the technologies needed to capture and distribute this supply (and inequalities within as well as between countries may still occur).

The first chapter, 'Solar Renaissance: Through the burning-glass', considers the time when solar technologies and stories started to proliferate across Western Europe, powered by the mythic topos of Archimedes wielding a burning-glass at the siege of Syracuse. In the written accounts and public demonstrations of natural magic, burning-mirrors and lenses were, therefore, simultaneously reconstructive and speculative – attempting to emulate and surpass their ancient inspiration, but in order to produce entertaining, uncanny and politically symbolic effects as well as, or perhaps instead of, knowledge. Assimilated into the solar iconography of European monarchies, these technologies supported the divine right of kings, and Neoplatonic and alchemical ideas. Yet burning-mirrors and lenses were also exploited in more subversive, often satirical, ways, targeting those same institutions, ideologies and philosophies. The handheld burning-glass, meanwhile, became a supreme simile within religious and secular writing, expressing the refractive index of faith and passion, and generating ideas regarding the self's relationship with the newly heliocentric cosmos.

It is perhaps one of history's many ironies that the most intense interest in burning-glasses in Northern Europe occurred during a period of decreased solar activity (the Maunder minimum), 1645–1715. Chapter 2, 'Bundling up the sunbeams: Burning into the Enlightenment', considers how, during the late seventeenth and eighteenth centuries, large burning-mirrors and lenses helped the emerging culture of natural philosophy to reflect upon the achievements of enlightened knowledge. The construction and artistic representation of these solar technologies spanned philosophical, professional, social and geographical boundaries, and fused conceptions of the ancient and the modern. Thermal optics became part of the appareil and iconography of the 'chemical revolution', facilitating important discoveries by Joseph Priestley, Antoine Lavoisier and others, and shaping public perception of the chemist. Particularly in the hands of satirists, the burning-glass was a brilliant vehicle for social and political comment, bringing to light – and fantasizing the photothermic annihilation of – error and injustice.

The turn of the nineteenth century was a time of political and social upheaval, and Chapter 3, 'Feeling the Promethean heat: Romantic radiance and the power of invisible light', explores how instrumentalized notions of the Sun's power were used to think through issues of liberty, imagination and environmental sensibility. These debates coincided with both William Herschel's discovery of 'invisible light' (infrared radiation) and a reinvigorated interest in mythological characters and narratives associated with the Sun. Solar power and its Promethean wielders (both real and fictional) were instilled with revolutionary potential and heroic individualism, with solar combustion framed as a liberating and revelatory process. The Romantics also revitalized the trope of the burning-glass, becoming a significant metaphor of the imagination within a wider aesthetic shift to expressive conceptions of mind. Chapter 4, 'A time of "Solidified Sunshine": Victorian imaginaries of solar energy', considers the paradoxical centrality of solar energy to nineteenth-century Britain, where modernity was built out of coal, a resource identified as 'solidified sunshine'. New optical and energy technologies were framed as attempts to emulate the Sun, while the developing field of thermodynamics sought to understand the source and duration of solar radiance. Previous scholarship has explored cultural anxieties about the 'death of the Sun' from mid-century. This chapter instead seeks to show how interest in utilizing solar power was, to some extent, reborn at this moment, coinciding with the discovery of photoelectricity and the birth of science fiction, influencing wider debates about the resources and social conditions of modernity, and humanity's relationship with the environment.

The fifth chapter, 'Bright futures: Solar science fiction takes off', explores the role of solar tropes in shaping science fiction (SF) as a genre from the early twentieth century, and the two-way traffic between SF solar energy narratives and scientific ideas and writings. In the works of Hugo Gernsback, Robert Heinlein, Isaac Asimov and others, the industrial and cosmic capture of solar energy functions as a synecdoche for technologically driven human futures. The chapter focuses, in particular, on the concept of the space solar power station (SSPS), which from the 1920s provided both utopian energy scenarios and disaster narratives for a genre growing in popularity, and reflected anxieties about nationhood, national resources, and the place of science and technology within society. Looking beyond those initial stories, the chapter also shows how the SSPS idea was taken up within both 'serious' scientific papers and SF thrillers of the 1960s and 1970s, gaining new impetus from developments in solar cell technologies, and also driven by concerns about fossil fuel depletion and environmental crisis. Chapter 6, 'Dark mirrors: Solar reflections in the nuclear age', also looks across the twentieth century, but traces solar energy's deployment as a 'death ray' within narratives of mass destruction, particularly as a metaphor for nuclear power. The development of the atomic bomb was figured frequently as a bringing of the Sun down to Earth, and stories of solar-fuelled violence (which drew upon older, photothermic conceptions of solar technology) enabled authors to interrogate the ethics of science and technology at a time when faith in scientific progress had been severely shaken. The chapter argues that even after the invention of the silicon photovoltaic cell, and the improved prospect of more generative visions of the Sun, the solar imaginary remains haunted by its deathly, but nihilistically attractive, mirror image.

The irony of solar energy as a Janus-like agent of salvation or destruction (of sustainability or climate change) is also present in the final chapter, 'Self-Renewable: The satire and psycho-thermodynamics of solar', which examines solar culture over the past forty years or so, during which time the necessity of a global transition away from fossil fuels has become acute. It focuses, in particular, on creative works that have considered questions of agency, both scientific and artistic, individual and communal, in the face of climate and energy crisis. Two main preoccupations are identified within these works: the psychology of energy (what we might call 'psycho-thermodynamics') and the use of the satiric mode. The chapter, therefore, explores both the influence of psychological and affective responses to energy forms and technologies upon our choices and behaviours, and the appeal of satire and irony when dealing with matters of global magnitude within a cynical zeitgeist that has a tendency to be suspicious of science and other forms of expertise. Is 'saving the planet' an inherently dull aspiration compared to the thrills of consumption? Does our sublime response to techno-scientific solutions to power generation and/or climate change help or hinder us? Is the story of our salvation via solar power as much of a myth as Archimedes's burning-glass? As an implicit reaction to these macro-scale issues, recent poetry has emphasized the individual, communal, domestic, everyday benefits of sustainable futures.

Understanding and utilizing the Sun's energy has been considered one of the ultimate goals of science and technology, and some of the world's most brilliant minds have dedicated themselves to this prospect. Yet solar energy translates into many different kinds of power, not just physical but also political and cultural. Our experience of it, therefore, cannot be reduced merely to frequencies and wavelengths, numbers and equations, matters of fact and dollar signs. Across the centuries, the literature of solar energy has tried to articulate and codify these other meanings, and has encouraged readers to reflect upon humanity's ecological status as both a product and an agent of the Sun's power. Like a kind of renewable itself, the solar imaginary is regenerated by the new environments and contexts into which it is transported – inviting new users to draw upon it afresh, and radiating its influence across all spheres of being.

1

Solar Renaissance:
Through the burning-glass

Archimedean dreams

We begin with a fragment of a dream, recounted from the night of 10 October 1910:

> *I was once more working at chemistry in the University laboratory. Hofrat L. invited me to come somewhere and walked in front of me along the corridor, holding a lamp or some other instrument before him in his uplifted hand and with his head stretched forward … with a clear-sighted (? far-sighted) look about him. Then we crossed an open space …*[1]

The dreamer was none other than Sigmund Freud (1856–1939), who sought to determine the causes and significance of this vision. Its 'most outstanding point' was the way in which his superior held the 'lamp (or magnifying glass) before him, with his eyes peering into the distance'. Freud 'knew at once' this was a substitute for 'someone greater': 'Archimedes, whose statue stands near the Fountain of Arethusa at Syracuse in that very attitude, holding up his burning-glass and peering out towards the besieging army of the Romans'. Freud had laid eyes upon this statue twenty-three days ago. Yet the principal cause of the dream was the news Freud had received that dream-day – his lecture theatre at the clinic was to relocate – reminding him of the diffident response he received from those above at the start of his academic career, when he had no teaching room to call his own. In the dream, therefore, Freud's wish had been fulfilled: the dean who ignored his requests years before was, in the form of Archimedes, finally satisfying his obligations to give him a 'ποῦ στῶ' (footing). This ancient genius and scientific pioneer, literally shining light into intellectual darkness, was showing Freud the way: Could any lesser thinker be worthy of the task? No wonder Freud admitted that these 'dream-thoughts were not exactly free from ideas of … self-importance' (*Interpretation*, p. 252n2).

Over the centuries, many thinkers, scientists, writers and artists have shared the powerful impression left by this image, striding confidently within the 'footing' it provided, including in the chemistry lab. Despite the implied illuminatory and/or ocular usage for the lamp or magnifying glass ('lupe') held by the professor, the burning-mirror or glass ('Brennspiegel') in Archimedes's possession had a darker purpose: to focus the Sun's rays upon the approaching Roman fleet, bringing about its pyrognostic destruction. For the pre-industrial age, the burning-glass (a term used interchangeably for mirror or lens) was perhaps the most evocative symbol of the absolute, almost divine power that could be unleashed by the application of knowledge, and had a fervent grip upon the early modern imagination.

Recent attempts to prove whether Archimedes constructed such optical devices have not been encouraging.[2] The credibility of the story is of little concern to this study. What is crucial is the imaginative agency of this narrative in inspiring not only writers and artists but also natural philosophers, instrument-makers and artisans. Indeed, because the story is so doubtful and mysterious – requiring the exercise of scientific, technological and philological expertise to negotiate its many challenges – it has captivated the mind and hand for centuries. In particular, during an early modern period susceptible to the attractions of natural magic, Archimedes's mirror offered a compelling story of knowledge facilitating power over nature, since it had secured control over the supreme source of light, heat and life. This episode of technological mastery was even figured as a kind of apotheosis, with Archimedes 'transform[ing] himselfe into *Iupiter*; thundering downe … lightning'.[3] The chapter will, therefore, begin by tracing the life of this myth into the heart of the Renaissance, allowing us to see how ambitions to collect and use solar energy reflect the intellectual, political and social aims of natural magic and science at this time. Later, the chapter will explore early modern relationships between material culture and the imagination, including the technologization of politics and theology, and its reflection in fantastical and satiric discourses.

As one would expect from such an evocative story of scientific genius standing up to military might, there are many accounts of Archimedes's attempts to defend his home city while besieged by Marcellus's Roman fleet (214–212 BC) during the Second Punic War. Several mention the use of an optical device – a six-sided mirror, according to Diodorus (c.90–c.30 BC).[4] None of the earliest sources (Polybius, Livy and Plutarch) describe the use of burning-mirrors, or even fire more generally.[5] However, in the late second century, Lucian (AD c.120–c.200) claims that Archimedes used 'his science' to burn the ships, while Galen (AD 129–c.200) notes the application of fire, either pots of flammable material or optics.[6] By the sixth century, Anthemius of Tralles (c.474–c.558) refers to the 'unanimous tradition that Archimedes used burning mirrors'; and Ibn Sahl (c.940–1000) and Ibn al-Haytham (c.965–c.1041) each have the Sicilian polymath in mind when they write their pioneering optical treatises.[7] During the Renaissance, accounts of Archimedes's achievements disseminated widely, and he became the role model for both the aspiring thinker and the inventor, at a time when the mechanical arts increased in their cultural value, partly because of such Classical *exempla*. The thermal optical work of other ancient figures (including Apollonius, AD c.15–c.100) was known, but various authors celebrated Archimedes particularly because he put his mathematics into practice, using '*arte perspectiue*' to create fire.[8] His ingenuity and military resistance elevated him to heroic status, exemplified by the topos of wielding the Sun's power in battle.[9] From the fifteenth century, technological developments allowed inventors to emulate Archimedes and find practical uses for optical innovations. For instance, Leonardo da Vinci (1452–1519) made numerous drawings of machines designed to grind metal concave mirrors for burning, not imaging, properties, such as soldering together parts of large sculptures.[10] At the same time, burning-mirrors seemed to provide Leonardo with answers to his theoretical enquiries about the Sun's heat.[11]

However, it was in the early modern books of 'secrets' (*libri secretorum*) that the *imaginative* possibilities of the burning-glass began to be emphasized. These technical books emerged from within the tradition of natural magic, and demonstrated a fascination with the power of technology over nature. Despite presenting themselves as 'how to' guides which explained how to produce extraordinary effects in a natural way, they also thrived on fantasizing tantalizing glimpses of wondrous, but apparently credible machines – perhaps as a way of seeking preferment from a courtly audience. The relationship between natural magic and natural philosophy was, therefore, a complex one, with books of secrets arguably contributing to the foundations of the 'Scientific Revolution', helping experiment become more communal, and inaugurating a popular image of science as a 'hunt after the secrets of nature'.[12] Solar technologies seemed to play a special role within these changes because of the established and increasing importance accorded the Sun in different but coalescing communities, genres and disciplines (including astronomy, alchemy and metaphysics).

The most famous of the *libri secretorum* was the compendious *Natural Magick* (1558) by the Neapolitan aristocrat and dramatist Giambattista della Porta (1535?–1615), a book revised in 1589 to include the developing cultural sensation of optical technologies. Although there were many mechanical marvels which demonstrated the wonder of nature, arguably nothing came closer than solar technologies in conveying the sublime power that humans could harness through practical endeavour, since 'catoptrick-glasses … shine amongst Geometrical instruments, for Ingenuity, Wonder, and Profit'.[13] Della Porta offered applications of an incendiary kind: one could use them to 'blow up Towers' by laying gunpowder and preparing one's glass at night ready to focus the Sun's rays the next day (p. 362). Despite his own grand claims, della Porta was suspicious of those made about Archimedes. Nevertheless, he made sure to boast of having designed a glass which 'exceeds the invention of all the Antients', by burning at an infinite distance (p. 375). For della Porta, this 'Artifice' can be used to 'inscrib[e] letters' in the sky 'as far as the Moon' (p. 375), inaugurating a trope which associated the act of writing with the action of the burning-mirror.[14]

In contrast to della Porta, the story of Archimedes was central to the optical work of the Jesuit natural magician Athanasius Kircher (1602?–80), professor at the Collegium Romanum. Kircher was not the first, nor the last, of the Jesuits to be interested in thermal optics.[15] Indeed, in the church dedicated to the Jesuits' founder, St. Ignatius of Loyola, the monk Andrea Pozzo's ceiling fresco (1691–4) depicts Ignatius's glorification, and the rays emanating at this moment are reflected in an angel's burning-mirror to form the Sun and Christogram, the symbol of the Society of Jesus.[16] But Kircher took the imaginative and intellectual investment in solar technologies to a new level of intensity and sophistication. Despite the importance Kircher conferred upon the Sun's light, however, his Jesuit affiliation demanded at least *public* loyalty to the Aristotelian, geocentric cosmos.[17]

Kircher's work on burning-mirrors was a kind of reconstructive archaeology, and included measuring the harbour at Syracuse itself (see Figure 1). Yet it

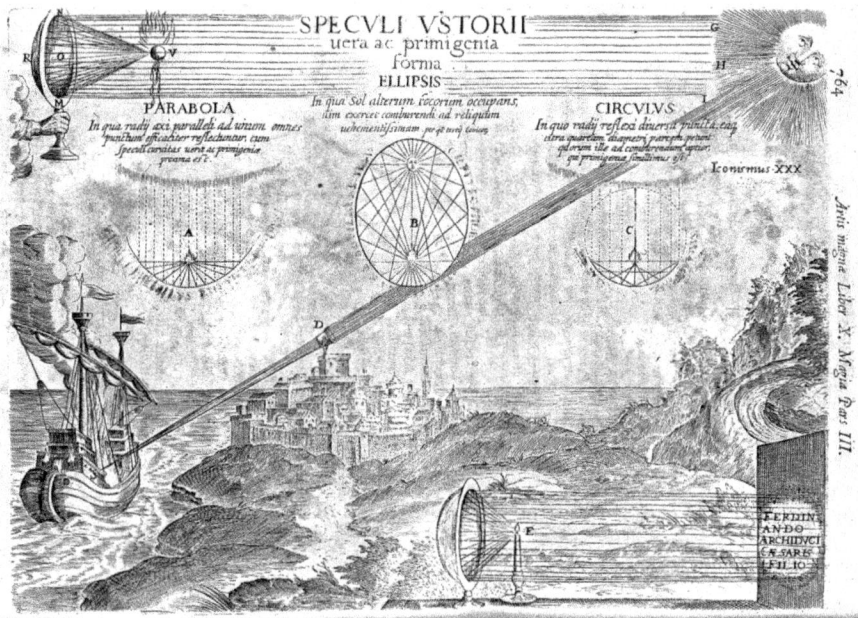

Figure 1 Athanasius Kircher, *Ars Magna Lucis et Umbræ*, second edn (Amsterdam, 1671), Plate XXX, facing p. 764. Wellcome Collection, CC BY.

was also innovative, demonstrating that increasing the number of plates upon which rays are reflected intensifies the radiation at the focus, suggesting that a number of smaller plane mirrors would function better than a single large parabolic, hyperbolic or elliptical one.[18] He also showed that a parabolic was more efficient than a spherical mirror, because it concentrated its rays over a narrower area.[19] Kircher published this work in *Ars Magna Lucis et Umbræ* (Great Art of Light and Shadow) (1646), which outlined many other optical wonders, including the camera obscura and, in the second edition (1671), the magic lantern. Especially following Kircher, the primal moment of Archimedes's mirror harnessing the Sun's power shone across the next century as a form of philosophical and cultural inspiration, establishing an interplay between knowledge of thermal optics as both the mastery of past narrative and the projection of future progress.

Fittingly, the frontispiece to the second edition is a cornucopia of natural, artificial and divine Suns, illustrating Kircher's metaphysics of light, which combined hermetic, Neoplatonic and cabbalistic ideas (Figure 2).[20] At the top is a glimpse of the divine light, the bottom of a solar disc marked with the

Figure 2 Athanasius Kircher, *Ars Magna Lucis et Umbræ*, second edn (Amsterdam, 1671), frontispiece. Wellcome Collection CC BY.

tetragrammon ('YHWH'), whose emanations pierce a ring of clouds populated by the angelic host. This light hits the Book of Scripture directly, imprinting its power through the Word, and embodying 'Auctoritas Sacra' (Sacred Authority). These divine rays inspire the (Platonic) extramissive production of light from the eye of human reason ('Ratio'), which seems to burn itself onto the page of another book of knowledge.[21] Below these divine and divinely inspired Suns sits the physical Sun, personified as Apollo. With its own, independent luminescence ('lux primigenia'), this material Sun produces three main beams: one reflected by the mirror of the Moon (itself figured as a technology emulating the Sun), held by Diana; the second refracted through a telescope to produce 'Sensus' (knowledge of the sensible); and the third penetrating into subterranean depths, and reflected by a mirror onto the back of the cave. In stark contrast, 'Auctoritas Profana' (Worldly Authority) is illuminated merely by candlelight. As critics point out, the frontispiece allegorizes kinds of human understanding as processes of vision.[22] Yet even more acutely, the illustration emphasizes light as not only an ocular but also a productive force. In both metaphorical and literal ways, it is able to effect material change, particularly upon the page: Kircher's book is itself testament to that.

The eagle and peacock were traditional symbols of the discipline of optics. Here, their two-headed equivalents glorify Kircher's patrons – the Habsburgs (particularly Archduke Ferdinand) – but they are also suggestive of the illustration's wider theme of doubling. Catoptrics is built into the structure of the frontispiece: divine and physical Suns, Sun and Moon, day and night, and the radiant garden and the shadow of the cave – the four sources of authority mirroring each of the others to create multiple reflections. The illustration shows us that the secondary light (lumen) of the physical world is a *reflection* of the divine, absolute light (lux) of God (or the Platonic Good), but never a true substitute. Nevertheless, the frontispiece implies that optical study aids the Christian in their understanding of God. In works of imagination throughout the period, the burning-glass conveys a sense of sublime experience which keep channels open to the secrets of natural magic and contact with the divine.

Alchemies of the Sun

The complex metaphysics of light can be traced to Plato's *Republic* (c.380 BC), which analogizes the intellectual light of 'the Good' and the physical light of the Sun. Neoplatonists saw this as more than an analogy, but as a real relationship.[23] Hence physical mastery over light also signified intellectual and spiritual enlightenment, as Kircher's frontispiece depicted in vivid fashion. Under the influence of Neoplatonism, from around the mid-sixteenth century Renaissance alchemists such as Heinrich Khunrath (c.1560–1605) and Robert Fludd (1574–1637) came to recognize the incendiary and generative powers of the Sun as imparting alchemical virtue to earthly materials.[24] For some, a special burning-glass might be the elusive philosopher's stone itself.[25] These ideas are evident especially in the work of the philosopher, astrologer and adviser to Elizabeth I, John Dee (1527–1609), famous for engaging in catoptromancy (divination by mirrors and crystals). As well as trying to intercept angelic images and messages, he was fascinated with catoptrical geometry, and recommended using mirrors as a way of accessing celestial energy. In Dee's natural astrology, heavenly bodies cause and order all motion in the terrestrial world through the 'radiated sympathy' of their light.[26] The principal celestial object is the Sun, an 'immense source of heavenly light' which 'poureth down his force', making the Earth yield its precious metals via a kind of sexual embrace.[27] The Sun's transmutational importance was apparent, therefore, in the 'laws of Anatomical Magic' that linked 'sun, gold, and man's heart' as celestial, terrestrial, and microcosmic reflections of each other (*JDA*, p. 161; Aphorism LXXIII).

The power of light, the 'first of *Gods Creatures*',[28] served as both metonym and metaphor for the operation of astral virtues emanating from all celestial objects. Dee saw these virtues propagating in the form of rays as light does, and, therefore, mathematical method could apply to natural astrology: the strength of influence of astral virtues could be calculated from the size, relative distance and angle of incidence of the celestial body from the Earth. Heat served as both the perfect analogue for, and indicator of, these virtues. Dee assigned the value of 100 to the Sun's 'maximal heat' when closest to the Earth, and when the axis of its cone of rays is perpendicular to a given point. From that baseline, the relative intensity of the Sun in other positions, and of other planets and stars, could be measured, and their astral influence determined.[29]

In the mid-1550s, Dee familiarized himself with many Classical, medieval-Arabic and recent works on practical optics, particularly burning-mirrors.[30] It was, therefore, a natural continuation of Dee's optical astrology to explore how the astral virtues he thought were cascading down to Earth could be imprinted upon earthly matter. Based on his reading of Roger Bacon and others, Dee seems to assume that parabolic mirrors would be the best surface upon which to focus astral rays and thereby achieve greater intensity than 'nature itself' is able.[31] In *Monas Hieroglyphica* (1564), Dee suggested dispensing with physical catoptrics altogether and instead forming a 'spiritual mirror' through the emblematic power of the 'Monas'. This was an esoteric glyph that combined astronomical symbols for Mercury, the Sun and Moon, and Aries (associated with the element of fire), and could enact a kind of transmutation by itself, forming a mirror which 'can reduce any stones or any metal to ... impalpable powders'.[32] A sonnet attributed to Dee describes how the ritual of the Monas will sacrificially 'Cutte the poudred Sonne in twayne'.[33] Presumably, this is an operation upon gold – the terrestrial 'spirit of Sol'[34] – and the linguistic equivocation between 'Sonne' and 'Sunne' emphasizes its potential Christian mysticism. Up in the heavens, both the Father and the celestial light were all-powerful and invulnerable, yet on Earth the Sons/Suns were sacrifices enabling human redemption and transcendence.

The Monas is one of many of Dee's ideas mocked by Ben Jonson (1572–1637) in *The Alchemist* (first performed 1610), characteristic of the comedy's satirical conflations of alchemical ambition with material desire and consumption. Dee's hieroglyph is reduced, appropriately, to a mere Blackfriars shop sign.[35] The play contains another – but neglected – allusion to Dee's catoptrics, when the duped Epicure Mammon believes that the con-artist Face is 'a divine instructor!', 'courted above Kelley', who can:

> call all
> The virtues and the miracles of the sun
> Into a temperate furnace.[36]

From around 1582, Dee's scrying companion was Edward Kelley (1555–97). As part of his patter to Mammon, the charlatan Face adopts Dee's audacious solar rhetoric; Jonson ridicules not only the blasphemous assertion that the Sun contains 'miracles' reproducible terrestrially but also the notion that anyone can 'call' forth such powers. The Monas was looked upon more favourably by others, including Kircher, who adopted the symbol himself.[37]

The practical experience of the burning-glass as a chemical catalyst is particularly prominent within the work of a later court alchemist, Nicaise Le Févre (c.1610–69). In 1652, Le Févre became a royal apothecary to Louis XIV, and at the Restoration was appointed Charles II's professor of chemistry and later apothecary-in-ordinary. The English king held the pursuit of alchemy in particular esteem, and granted Le Févre space for a laboratory at St. James's Palace. Le Févre's intellectual reputation was based principally on *Traicté de la chymie* (1660), translated into English as *A Compendious Body of Chymistry* (1662) and subsequently retitled *A Compleat Body of Chymistry* (1664). His treatise includes a description of the '*Solar Calcination* of Antimony', which produces the white powder antimony trioxide. Unaware of the oxidation process which 'encreases the substance of Antimony', Le Févre sees this experiment as confirming light's generative properties: 'that *Promethean* Fire, which gives life to all natural bodies'. Antimony is particularly receptive because it 'hath a kind of natural Magnet in it self' which attracts 'this noble kin and similar light'.[38]

Le Févre's elevation of light would also grace his greatest work for Charles, who seemed to have employed him almost exclusively in order to recreate the fabled immortalizing 'cordial' of Sir Walter Raleigh, as a centrepiece of the monarch's rumoured utopian ambitions. Le Févre's resulting *Discourse* presented the 'Sovereign Remedy' as a reflection of Charles's restoration of the state.[39] The medicine's ingredients (including gold, pearls and spices) were described as natural burning-glasses: 'filled with a concentred Light, which can remedy all [the population's] evils' (p. 97). In particular, gold as the 'King of Metalls' had a 'harmonical relation not onely with the Celestial Sun of the Great World, but also … with the Sun of the Little world, which is the Heart of Man' (p. 59), restating – with an alchemical inflection – William Harvey's solar metaphysic of monarchical power (and perhaps intentionally echoing Dee's 'Anatomical Magic').[40] The famous intelligencer Samuel Hartlib (c.1600–62), seeming to conflate the two solar aspects of Le Févre's work, remarked in his notebooks that burning-glasses could calcinate antimony to produce 'a soveraigne Medecin' very different to the 'hurtful' substance acquired via furnace heat.[41] Not all were taken in by the claims made about Le Févre's medicine, however, and Samuel Butler (bap.1613, d.1680) – a satirist known for his rationalist critiques of occult and natural philosophies alike – included 'a powder of the sun, / By which all doctors should b'undone' in his poetic list of the excesses of the age.[42]

More sincerely, Hartlib's acquaintance, the poet and essayist John Hall (1627?–56), saw the burning-glass itself as a sublime product of alchemical union:

> Strange Chimistry! can dust and sand produce
> So pure a body, and Diaphanous;
> Strange kind of Courtship! that the amorous Sun
> T'embrace a Min'rall twists his rayes in one,
> Talk of the heav'ns mockt by a sphear, alas
> The Sun it self's here in a piece of glasse:[43]

'A Burning-glasse' (1646) gestures towards many ways in which new optical technologies were marvelling the early modern audience. The small, handheld burning-glass became, by at least the mid-seventeenth century, both a wondrous tool in experimental philosophy and a fashionable accessory within polite culture. In a courtship trope, Hall captures the uncanniness of the burning-glass: transparent in form, and yet opaque and mysterious in its refractive ability. The Sun 'it self' in a 'piece of glasse' is a hyperbolic image typical of Hall's metaphysical trickery, and follows the natural magic tradition in seeking emblematic images of nature within the productions of man. However, Hall's poem does not eulogize the inventors and craftsmen who manipulate the raw materials of soda, lime and silica. Instead, in accordance with the erotic potential of the central conceit that 'A Mistris eye is but Loves Burning-glasse' (l. 34) in focusing the emotions and creating a kind of spark, the poem grants the substances themselves the agency to bring a new 'body' into being. In the poem's centre, the speaker imagines what the ancients could have achieved with such technology, including 'Had *Archimedes* once but known this use / H' had burnt *Marcellus* from proud Syracuse' (ll. 21–2). This suggests Hall doubted the story was true, perhaps following René Descartes (1596–1650), who in his *Dioptrique* (1637) contested the myth by claiming that a mirror with a long focal length, and hence large focused spot, could not heat more strongly than the Sun's direct rays.[44] However, as we have already seen, in the same year Hall's *Poems* appeared, Kircher's *Ars Magna Lucis et Umbræ* produced a robust defence of Archimedes's invention.

Hall's poem invites the reader to attend to the materiality of the burning-glass and, by corollary, to consider how a sensory encounter with a material object is conveyed within a mediating text. As numerous studies have shown, glass and other reflective or refractive surfaces have long been a source of literary fascination, providing a wide range of figurations. They can, for instance, apparently capture images which take on a life of their own; they can reflect things as they are, and also show how things will, should (or should not) be.[45] Moreover, during the sixteenth and (especially) seventeenth centuries, the *scopic* possibilities (both literal and metaphorical) of glass came to prominence, in the form of the telescope and microscope. The burning-glass offered a different kind of ocularity, but connected metonymically to these newly discovered properties of glass. While the handheld burning-glass was becoming a quotidian object, the alchemical emphasis of Hall's poem suggests that he was also acknowledging and partly writing for a learned audience (such as the Hartlib circle) who would appreciate the metaphysical invention of his poem. The title – drawing upon a tradition of 'mirror' or 'glass' books envisaged as kinds of practical instrument[46] – implies that the poem is a veritable 'burning-glass' itself, focusing together fragmentary strands of knowledge and experience: a meta-textual metaphor that would be returned to by numerous authors over the centuries.

Burning news and parabolical fictions

In seeking to invest his rule with solar meaning, Charles II was drawing upon
the kind of commonplace heraldry of sixteenth- and seventeenth-century
European monarchy we have already witnessed in the frontispiece to Kircher's
Ars Magna Lucis et Umbræ. Images of the Sun's light pervade Elizabethan
art, as metaphors for the Golden Age and the virginity, purity and faith of
its monarch. The Stuart court propagated a solar iconography with new
emphases in Neoplatonist and hermetic emblematics, heliocentric cosmology
and optical technologies.[47] James VI (1566–1625), soon James I, asserted in
Basilikon Doron (1603), his treatise on government, that God uses the 'light
of nature' as the medium by which he 'imprint[s] in mens minds ... the loue
of all morall vertues', while also maintaining omniscience with his 'all-seeing
eye, and penetrant light'. Divinely anointed monarchs, therefore, should reflect
God's actions and abilities in order to fulfil their obligations to their subjects,
in exchange for absolute rule.[48] Jacobean art and culture rapidly assimilated
these Neoplatonic associations. Jonson's 'Panegyre' on James's entrance to his
first session of parliament, on 19 March 1603, celebrates the 'thousand radiant
lights' which extramissively 'stream' from his eyes through every 'nook' of
his new 'realm'.[49] Among many masques and processions which dramatized
monarchical powers of illumination, Jonson's *Hymenaei* (1605) includes a
figure of Truth whose 'right hand holds a sun with burning rays' who concedes
to the superior wisdom of James (the Sun in the hand being a common emblem
of the faculty of understanding).[50] Later, the architecture of Inigo Jones (1573–
1652), particularly the Roman temple-style portico added to the west face of St
Paul's Cathedral, sought to emphasize the radiation of Platonic light and virtue
through the macrocosm of the nation, and the frontispiece to the King James
Bible (1611) distinguished the Sun and the Moon as material reflections of the
divine light of Creation.[51]

The Stuart monarchy's image of divine right as manifested in Neoplatonic
solar symbolism also became the province of the emerging court engineers,
who were seeking to prove their practical credentials. They included the French
Huguenot Salomon de Caus (1576–1626), a garden designer and hydraulics
expert. De Caus's projects apparently incorporated mechanical illusions,
including an enormous fountain of Apollo and the Muses at Somerset House;
while for the heir to the throne, Henry, Prince of Wales (1594–1612), de Caus

designed a number of grottoes in the grounds of Richmond Palace.[52] Around 1614, de Caus was invited by James I's son-in-law, Frederick V (1596–1632), to design the Garden of the Palatinate in Heidelberg, where he installed a version of Memnon's statue, which appeared to speak when animated by sunlight.[53] Despite the sense of wonder they elicited, de Caus's solar projects paled into comparison against the visions of another artificer from the continent, Cornelis Drebbel (1572–1633). Born in Alkmaar (Holland), Drebbel came to London around 1605, and spent much of the rest of his life in England. Entering into the service of James I and particularly Prince Henry, Drebbel had a laboratory at Eltham Palace from at least 1610, where he entertained curious visitors with optical entertainments and other inventions. Drebbel's glass-grinding techniques became widely known throughout Europe, and he was lauded for his compound microscope and wonders such as the camera obscura and magic lantern. Drebbel claimed that his projections could change the appearance of his clothing, and make it appear as if ghosts were rising from the Earth.[54]

While it is unknown how many of the optical projects were realized, there is evidence that Drebbel was keen to extoll the virtues of solar power to his royal patrons, in discourse reminiscent of della Porta's incredible descriptions of 'secrets'. Indeed, unsure of his prospects upon the death of Prince Henry (November 1612), Drebbel wrote to the king seeking further patronage, tantalizing his majesty with the vision of a musical fountain depicting Neptune, the Tritons and sea-goddesses bathing, and Phoebus riding along in the sky. These animations

> shall be accomplished by the rays of the sun alone, … In case your Majesty might wish to feast your eyes … when the whole sky is clouded, then you will nevertheless be able to do this by simply touching a small glass vessel with a warm hand.[55]

What marvel was this? Not only did it conjure up a dance of the gods powered by the Sun or by haptic embrace but also offered the chance, if only for an instant, to emulate Apollo himself!

Drebbel continued in royal service, but it is unknown whether the promise of realizing this solar delight sustained the king's patronage specifically. However, this kind of scheme (along with those of de Caus) proved substantial comic fodder for Jonson in *The Entertainment at Britain's Burse*, staged 11 April

1609, to celebrate the opening of 'The Burse' or New Exchange, a shopping centre in the Strand, with attendees including members of the royal family. Jonson's masque – tonally more comic than satiric – both celebrates The Burse as satisfying his society's desire for luxury and rarity, and derides this elite, proto-consumerist impulse.[56] The Shop-Master's products include the latest instruments of natural knowledge and, particularly, optics – 'sundials, hourglasses, looking-glasses, burning-glasses, concave glasses, triangular glasses, convex glasses, crystal globes' – and as his sales pitch progresses he presents himself increasingly as an expert in mechanics, optics and engineering. Indeed, the Master compares his stock to the installations of de Caus and Drebbel: 'A statue too of Apollo here will do something. Now, if thou be'st the god of music, let's hear thee!' (an actor dressed as the statue then proceeds to sing). Drebbel, whom Jonson probably knew via Gresham College, is then the particular focus, with the Master setting himself up as a rival engineer: 'I would my antagonist at Eltham were here … This is past the heat of hands or the beams of the sun.'[57] Of course, the meta-theatrical joke is that the 'Apollo' is merely a man dressed up as an automaton, not a mechanical object of wonder. Were the claims of de Caus and Drebbel similarly overstated? Whose authority can be trusted in a commercial culture of developing expertise in new kinds of knowledge? Even if the projects of de Caus and Drebbel lived up to their assertions, the masque asks what the true value of such uncanny emulations of human or divine animation might be, and whether an actor performing as Apollo is any less entertaining or instructive than a feat of engineering doing the same.

Over a decade later, Jonson teased Drebbel again, in *The Masque of Augurs* (1622). His caricature 'Vangoose' is an Englishman seemingly faking a Dutch accent to bolster his kudos, a 'projector of masques' who claims to produce illusions by 'de catopricks'. After enticing the audience with boasts of creating wonders in 'dis very room', Vangoose pretends to summon an optical 'anti-masque' of 'deformed pilgrims' via an incantation of 'Hocus pocus'.[58] In a parodic display similar to *Britain's Burse*, Vangoose calls 'Apollo' and 'Jove' to the stage: burlesquing Drebbel's optical entertainments, these gods are portrayed by flesh and blood, not tricks of the light. In *Augurs*, however, the royal family are not only in the audience, to be addressed as potential investors in Vangoose's schemes, but also involved in the performance, with Prince Charles (by then, heir to the throne) as one of the masquers (presumably Apollo or Jove), not just laughing at but telling the joke.[59] The Prince's carnivalesque participation is perhaps indicative of the early Stuart court's attitude to its Neoplatonic, solar iconography: receptive to its propagation, representational benefits and artistic possibilities, but suspicious – and playfully subversive – of its serious hermetic value. For Jonson, the mechanical marvels of de Caus and Drebbel – promising much, but either unfulfilled or grossly expensive – were certainly the stuff of comedy, but perhaps also an iconographical step too far, highlighting the artificiality and mercenary potential that surrounded this metaphysical cult.

Drebbel's proposals to Prince Charles were even more ambitious. According to the French astronomer Nicolas Claude Fabri de Peiresc (1580– 1637), Drebbel had been inspired by the New River project (1608–13) of Sir Hugh Myddelton (1560–1631), which included piping water to individual houses, and suggested supplying heat for culinary fires in a similar way. On a hill just outside London, solar mirrors would ignite some sort of non-consumable substance whose heat would be transferred by pipe to the capital, a method resembling the Roman hypocaust. Nothing else is known of the scheme – perhaps because Drebbel was asking for funds of 20,000 pounds – and it is unclear how much of the tale had been constructed by de Peiresc or his source, Drebbel's son-in-law Abraham Kuffler (1598–1657).[60] Notwithstanding the gross impracticality of this project, it confirms the openness of the early modern imagination to the possibilities of a solar power that could transform domestic life.

Amid the Shop-Master's many audacious claims in *Britain's Burse*, one is at the zenith of optical aspirations: 'I am promised a glass shortly from a great master in the catoptrics, that I shall stand with o'th' top of Paul's, when the new spire is built, and set fire on a ship twenty leagues at sea in what line I will by parabolical fiction' (ll. 145–7; *Jonson*, III, 364). The old St Paul's Cathedral had been destroyed by fire in 1561. By 1609, the spire remained unbuilt and might have been a standing joke, with some added topicality, since Inigo Jones had recently made proposals for its reconstruction.[61] The situational irony (and sense of an accident waiting to happen) of placing a burning-mirror atop a building previously destroyed by fire is lost in the Master's enthusiasm, and his plan would be taking the militaristic ambitions of the church to a new level. The pun on 'parabolical fiction'/parabolical section[62] highlights how the Master's claims, and those made of burning-mirrors more generally, challenge scientific plausibility.

In Jonson's *The Staple of News* (1626), a burning-glass is proffered mockingly as part of the *continental* military machine, to similarly incredulous effect. The 'staple' is a kind of news agency, where clerks divulge to visitors the latest gossip, in part parodying the far-fetched stories found in contemporary *coranto* news-sheets.[63] The play suggests both a fascination with and a suspicion of the growth in news media, but there has been some dispute over the objects and severity of the satire.[64] Nevertheless, critics agree that Act 3 Scene 2 – where we see how the staple trades in nationalist propaganda, conspiracy theories and fantastical stories of invention – is the play's key satiric scene. Here, the clerk Tom Barber reads out stories acquired from Catholic sources on the continent for the visiting Pennyboy Junior, who is keen to hear of developments in the Thirty Years' War. Barber reports that under a new leader the Jesuits have relinquished their vows, to become 'the only engineers of Christendom', supplying inventions which, to varying degrees, push the limits of plausibility. The device that Pennyboy finds most worrying comes in news from Florence:

Thomas	They write was found in Galileo's study
	A burning glass, which they have sent him, too,
	To fire any fleet that's out at sea –
Cymbal	By moonshine, is't not so?
Thomas	Yes, sir, i'the water.
Pennyboy Jr	His strengths will be unresistible, if this hold! (3.2.53–8; *Jonson*, VI, 85)

This parodic report is dubious but within proximity of the plausible. Galileo (1564–1642), alluded to directly by Jonson on this single occasion, was famous for his expertise with the telescope, but not known for working on thermal optics (although he had made reference to the story of Archimedes).[65] Similarly, although Galileo had some Jesuit supporters, in the 1620s he was embroiled in a dispute with Father Orazio Grassi (1583–1654), Professor of Mathematics at the Jesuit College in Rome.[66] 'Moonshine in the water' was proverbial, a reference to something trivial and phoney,[67] and Cymbal's suggestion that Galileo is using the Moon's rather than the Sun's rays intensifies the comedy of Pennyboy's gullibility. Pennyboy's hopes are allayed slightly when he hears that Protestant Europe has sublime military technologies of its own, including the Dutch 'invisible eel' designed by 'one Corneliuson', which can 'sink all / The shipping' at Dunkirk (3.2.58–61), exaggerating the abilities of the rudimentary submarine Drebbel had tested along the Thames in 1621.[68]

Jonson would have read of the particular fascination of Archimedes's burning-mirror to mathematicians and engineers within his copy of Dee's 'Mathematicall Preface' appended to Euclid's *Geometry* (1570).[69] While both *Britain's Burse* and *The Staple of News* trade comically on the incredulity of Archimedes's endeavours (and their emulation), Jonson's mock-heroic weapons of mass destruction are only at one remove from the contemporaneous inventions he may have encountered. Via Gresham College, Jonson might have known the work of John Napier of Merchiston (1550–1617), who was not only the inventor of logarithms but also an expert in thermal optics.[70] In June 1596, Napier claimed to have 'the invention, proofe and perfect demonstration, geometricall and alegebricall, of a burning mirrour' which unites the Sun's beams in 'one mathematicall point' (actually a parabolic mirror's focus images the Sun as a disc). Far from keeping his invention in the world of theory, he asserted its military application and Archimedean heritage: 'This invention serveth for burning of the enemies shipps at whatsoever appointed distance.'[71] A fervent anti-papist, Napier saw his mirrors as a way to see off Spanish invasion, at a time of undeclared, but real, war (1585–1604). Napier, therefore, would make an ideal candidate for the anonymous 'great master in the catoptrics' lauded by the Shop-Master in *Britain's Burse*, fanatically wishing death and destruction upon his religious enemies atop St Paul's.[72]

The Staple of News might have had a more topical satiric focus, since in the year it appeared (1626), Jonson's friend and patron William Drummond of Hawthornden (1585–1649) sought the crown's approval for a patent to manufacture fourteen inventions 'of use and profit to the State'. They included thermal optical devices which, in their della Porta-like claims, sounded as implausible as those reportedly found in Galileo's study:

> Burning Glasses … by which, at whatever distance, whether on sea or land, any combustible stuffs … may be set on fire. All these, though consisting of glasses shaped of various conic sections, concave and convex, and of other curved surfaces, … and burning by reflections as well as by refraction … will be called *Glasses of Archimedes*.[73]

Drummond is insistent that his new armaments originate in 'the application of mathematical and physical principles' (p. 157), and placing his own work in the Archimedean tradition reinforces this claim. Unfortunately, we know little about the instrument itself because the only surviving evidence is Drummond's application. It is also difficult to ascertain whether a satirical, or merely comic, gibe at his friend is intended. Nevertheless, Jonson's joke confirms the potent relational meaning that exists between burning-glasses and other forms of optical technology, illustrates the embeddedness of the burning-glass within the early modern cultural imagination and gives us an early occurrence of its satiric possibilities, especially in order to highlight the discrepancy between the expectations of technology and their real abilities. Despite Pennyboy Junior's worries, Galileo's *telescope* was changing the world more acutely than a lunar death ray could ever muster. Yet Jonson shows that even just *stories* of such devices commanded by either side, once filtered through the propaganda machine, have the potential to send opponents into a panic. These 'parabolical fictions', reflected by and between the cultures of the court, theatre, marketplace and laboratory, sometimes possessed an agency in the real world.

Shining heads and smoking hearts

Iconography was no match for military prowess during the Civil War, and Charles I was decapitated outside the Banqueting House of Whitehall Palace on 30 January 1649. By the next day, however, his head was blazing a new kind of power across the kingdom, as *Eikon Basilike* (The King's Image) was published, with an astounding thirty-five editions appearing that year.[74] Purportedly the work of Charles himself, it was later claimed to have been assembled by the king's chaplain, and future bishop of Worcester, Dr John Gauden (1599/1600?–62).[75] The book sought to vindicate the king's actions, and create an image of divine kingship and martyrdom that highlighted the fallen monarch's Christic and Davidic parallels.[76] Crucial to this strategy was the frontispiece engraved by William Marshall (*fl.*1617–49), but claimed to have been designed by Gauden (perhaps with Charles's help). The illustration converges a number of symbolically potent ideas about the king (Figure 3). Charles is shown at prayer before an open Bible, penitently holding his heart, and surrounded by three crowns: the vain crown of earthly success, cast aside on the floor; the crown of thorns, clutched in his right hand; and the radiant crown of heavenly glory, a vision of his future reward, connected to his left eye by a beam of light labelled '*Coeli Specto*' (I look at the Heavens).[77] Mirroring this beam on the left-hand side is a ray of light, annotated '*Clarior é tenebris*' (More

Figure 3 *Eikon Basilike: The Pourtraicture of His Sacred Majesty in His Solitudes and Sufferings* (London, 1648 (1649)), frontispiece. Courtesy of University of Liverpool Special Collections & Archive.

brightly out of darkness), which cuts through the clouds to form a focus at the back of the king's head, concentrating its power upon him. These geometrical lines of light-rays are reminiscent of those in hermetic illustrations, which often depict the propagation and reflection of the solar 'virtues' explored earlier in the chapter.[78] In the background, meanwhile, are numerous conceits appropriated from popular emblem books.[79] The embedded emblematic and religious images have been discussed as part of the 'complex history of the production and reception of images of power',[80] but surprisingly the optical dimension has not received similar treatment. Resituating the frontispiece within the context of solar technology and the material culture of the Stuart court can shed further light on its iconographic power.

The source of light that penetrates the clouds is obscured, but is at least metaphorically, if not literally, identified with the Sun and, by corollary, the Creator, thereby asserting the divinely ordained 'full lustre of Kingly Power'.[81] More specifically, the image renders pictorially Charles's appeals to the 'Sun of righteousnesse, thou sacred Fountaine of heavenly light and heat, [to] cleare and warme my heart', and 'Make … mine innocency to shine forth' (pp. 215, 136). The image of the Stuart monarch as Sun-god was culturally pervasive, as we have seen, but there was a strong popular tradition of identifying Charles (both positively and negatively) in this way,[82] and in deploying scientifically and pseudo-scientifically inflected variations upon this motif. In the astrological tradition, the Sun signified monarchy, and astrologers on both the Royalist and Parliamentarian sides sought to co-opt celestial phenomena to their causes. William Lilly, for instance, interpreted the appearance of three 'Mock-Suns' (Parhelia) on Charles's birthday in November 1644 as a portent of the king's defeat.[83] Moreover, the non-partisan Thomas Povey (1613/4–c.1705) saw the Long Parliament of 1640–1 as receiving 'enlivening Influence from the King', but then 'like a Burning-glasse contract[ing] the Sun beames into it selfe, … able to give fire to almost any Designe it pleased to reflect on'.[84]

The ray of light surrounding the crown of glory is perhaps the frontispiece's most significant interpretative difficulty. Is this object shining its light into Charles, or perceived via illumination by the monarch's eye, in accordance with the (waning, but still symbolically potent) extramission theory of sight? More likely is that the crown is itself a projection, a phantasm from within the king's mind of the divine blessing awaiting him, a refraction of the heavenly beam his head is receiving. Indeed, this vision of heavenly power resembles the kind of religiously inflected image (both divine and diabolical) that natural magicians projected via magic lanterns. While *Eikon Basilike* pre-dates Kircher's account in the second edition of *Ars Magna Lucis et Umbræ* (1671), the optical mechanism suggested in the frontispiece resembles the magic lantern more than the well-established camera obscura. One answer to this conundrum can be found within the Stuart court, which a couple of decades previously had voraciously consumed Drebbel's mechanical wonders. The extravagant performances of natural magic at Eltham Palace, therefore, lived on in an iconic image, bound up with the king.

While the magic lantern parallels are certainly present, the king's head is also a kind of burning-glass, focusing the light of righteousness it receives from on high to produce a radiant reflection of divine power. If the frontispiece was indeed the design of Gauden, we know that he employed burning-glass similes elsewhere.[85] Besides, as we will see below, such figurations were common within contemporary religious discourse, and included comparisons between Christ's redemptive power and the mediating capacity of the burning-glass.[86]

A day after the Royalist cause was thought to have shattered, therefore, the executed monarch was visually immortalized as a godly instrument via a developing iconography of thermal and projectile optics. The instrumental simile sought not to diminish the Christ-like king, but reassert his divinely sanctioned sovereignty in a display of sacred radiance.

During a period of expansion in the scopic imagination, theological writings asserted the many affinities between spiritual and optical mediation; but the burning-glass's acute metaphoricity enabled it to become perhaps the optical simile par excellence. For instance, drawing upon St Augustine's notion that man is 'enlightend' by God in the 'bedchamber of the heart', the pastor Anthony Burgess (d. 1664) compares 'godly assurance' to 'the burning Glasse that … cause[s] a fire to be kindled within', while for John Flavel (1630?–91), 'Faith is that burning-glass which contracts the beams of the grace, and love, and wisdom, and power of Jesus Christ together; reflects these on the heart, and makes it burn.'[87] One of its most interesting formulations, though, is in the work of the evangelical Anglican clergyman Anthony Horneck (1641–97). A major figure in London's reformation societies in the late seventeenth century, Horneck would be a significant influence upon John Wesley (1703–91) and the Methodist movement.[88] From 1669, he was also a fellow of the Royal Society, and would have been aware of developments in thermal optics: he was present, for instance, at a meeting when Robert Hooke (1635–1703) 'made a proposition of highly commendable improvement of … burning-glasses', and may well have witnessed such devices.[89]

The embeddedness of material culture, especially optics, is particularly acute within Horneck's *The Great Law of Consideration* (1678), in which he argues that one's spiritual self rests upon 'improving that rational Faculty, the great Architect hath bestowed'.[90] Microscopes, telescopes and other optical devices are brought forth as analogies regarding humanity's limited and/or warped spiritual vision.[91] However, the central simile for the meditative state of 'Consideration' itself is that 'inward thoughts, like Sun-beams in a Burning-glass, unite and continue so long upon these spiritual objects, till they set the heart on fire' (p. 12). The duration of the photothermic process is crucial to Horneck's argument: the spiritual insight cannot be produced instantaneously, but only with due concentration of the individual's thoughts. 'Consideration' invigorates internalized faith and allows for the conscience, seated in the heart, to receive the grace of God. 'Blessed Fire … purifies' the heart, 'elevates' the soul, 'destroys the dross, but preserves the gold, and burns away all unclean and inordinate passions' (pp. 310–11). Horneck's materializing account of this divine cleansing is loaded with alchemical connotations that combine Christian hermeticism with the kind of burning-glass trials he may have witnessed at the Royal Society.

The deepest resonance of Horneck's burning-glass image within his regime of spiritual instrumentalism relies upon not only the numerous associations of the Sun with God but also a cluster of ancient, medieval and early modern meanings surrounding the heart. Probably the most important is the Pauline aphorism that 'God … hath shined in our hearts' (II Corinthians 4:6), derived from Old Testament lore that addressed the heart as the place of spiritual and moral understanding. Classical theories also gave prominence to the heart both spiritually and physiologically, associating this organ with the generation of warmth. For Aristotle, the heart gave animation to the soul via its production of refined air, or '*pneuma*'; the Platonic and Galenic bodies were vitalized and regulated by the heart, considered a kind of furnace.[92] Harvey's discovery of the circulation of the blood formed the crux of the mechanistic physiology that overthrew humoral conceptions, but maintained the importance of the heart to both physical and spiritual vitality, at least for a time.[93] In the early modern period, therefore, spiritual and corporeal conceptions of the warm heart seemed to mutually sustain each other. Luther and Calvin would follow Paul in seeing the heart as the seat of the conscience, becoming a crucial element within the discourse of Protestant introspection and direct communication with the Almighty.[94] Most famously, John Wesley's 'Aldersgate experience' of evangelical conversion (at a Moravian meeting, on 24 May 1738) involved his heart being 'strangely warmed'.[95] The correspondence between helio- and cardiac-centred systems was expressed most fervently by Harvey (1578–1657), as he sought to vindicate his own discovery alongside Nicolaus Copernicus's similarly revolutionary findings: 'This organ deserves to be styled … the sun of our microcosm just as much as the sun deserves to be styled the heart of the world. … The heart is the tutelary deity of the body'.[96] For Horneck, the burning-glass was, therefore, an ideal simile through which he could construct an intellectualized and introspective religious identity, expressing a kind of refractive index of the soul (the microcosm of divine energy). Thermal optics, seeming to bridge literal and metaphorical uses of light, bringing two hearts together, offered a conceptual model for Horneck's method of spiritual illumination.

Later in the treatise, Horneck subjects the Sun itself to his meditative process. There is a kind of performativity about Horneck's argument at this point: he notes how 'Consideration can metamorphose objects, and spiritualize them', and then does so rhetorically within the passage itself, with the physico-theological analogy between Sun and God almost transforming into identity (precariously hovering above idolatrous sun-worship). The 'kindly beams' of this 'bright star' are borrowed from the 'shining rayes' of the 'increated Sun' (pp. 302–3), called the 'Father of lights' in Jas. 1.17. The conceit of God as 'increated Sun' is reprised in the book's frontispiece, which literalizes the figure of Consideration 'heat[ing] the heart within' to 'kindle the holy fire' (p. 13), and is accompanied by a Latin motto which translates as 'Consideration unifies, burns, illuminates, divides, attracts and renews' (Figure 4). This image concretizes evocatively Horneck's burning-glass simile, and externalizes the 'inward thoughts'. Perhaps to avoid accusations of idolatry, God the 'Father of Lights' is replaced with the personified Sun and intermediary angel, the heavenly host viewed traditionally as the mediators of God's message.

Figure 4 Anthony Horneck, *The Great Law of Consideration: Or, a Discourse*, second edn (London, 1678), frontispiece. Wellcome Collection CC BY.

Penitently kneeling, at the moment of 'Consideration' the human receives enlightenment from the illuminated angel, who administers the Sun's beams via a burning-glass. In so doing, the frontispiece seems to draw upon St Teresa of Avila's (1515–82) accounts of spiritual transverberation, of being 'utterly consumed by the great love of God' when pierced by an angel's golden spear, and 'utterly blinded' in the soul when struck by the 'Sun of Justice'.[97] The eyes and the heart become the sites of direct communication with this solar God. With a Calvinistic hand on the heart that is the focus of the divine light,[98] the human's gaze is guided by the hand of the angel to look up directly at the eyes of the personified Sun, while the angel's own pensive facial expression seems to parallel the concentrating power of the lens. The purifying effect of 'Consideration' is visualized as a cone of smoke from the meditator's breast.[99]

The heart warmed by a burning-glass was, therefore, a particularly potent image within the discourse of evangelical Christianity, and drew upon numerous traditional and emerging ideas that surrounded the body's central organ. These associations also stimulated many poetic explorations of Platonic love and romantic desire, offering ways to express the rebounding of desire, and reflection upon the inward motions of the self. For instance, Sir John Suckling (1609–41), infamous gamester, wit and courtier to Charles I, wrote a number of technologically inflected amorous poems, including the sonnet 'Love's Burning-glass' (c.1626–32):

> WONDERING long how I could harmless see
> Men gazing on those beams that fired me,
> At last I found, it was the Chrystal Love
> Before my heart, that did the heat improve:
> Which by contracting of those scatter'd rayes
> Into it self, did so produce my blaze.
> Now lighted by my Love, I see the same
> Beams dazle those, that me are wont t'inflame,
> And now I bless my Love, when I do think
> By how much I had rather burn then wink.
> But how much happier were it thus to burn,
> If I had liberty to choose my urn!
> But since those beams do promise only fire,
> This flame shall purge me of the dross, Desire.[100]

The source of the 'beams' is perhaps Suckling's cousin, Mary Cranfield, whom Suckling had imagined previously as a Sun enflaming his desire.[101]

The unattainable object of the sonnet, surrounded by a crowd of fleeting admirers, is dehumanized into a mass of 'scatter'd rayes'. The sonnet focuses on the 'Chrystal Love' that resides within, and is 'contract[ed] … / Into it self'. Instrumentalizing this love via analogy with the burning-glass, the speaker attempts to rationalize his feelings of the heart. The intensity of this 'Love' is attributed to his own imaginative capacity: the speaker acknowledges this, but chooses not to 'wink', and instead basks in the enflamed emotion. Yet this desire is only the 'dross': a deeper love resides beyond the visual 'beams', and will not be burnt away when passion fades. Suckling's conceit, therefore, is figured self-consciously as a kind of experiment upon the self, using the burning-glass metaphor to help him conceptualize his feelings.[102]

For the poet Katherine Philips (1632–64), meanwhile, the concentrating power of the burning-glass served as a fitting simile for the 'society of friendship' she established among her closest female companions:

> But as though through a burning-glass
> The Sun more vigorous doth pass,
> Yet still with general freedom shines;
> For that contracts, but not confines:
> So though by this her beams are fixèd here,
> Yet she diffuses Glory everywhere.
> Her mind is so entirely bright,
> The splendour would but wound our sight[.][103]

Philips's verse epistle challenges the conventional image of the female eye as the Sun. Here, it is the addressee Anne Owen's 'mind' which is 'entirely bright', concentrating its 'beams' of virtue within a welcoming community of like-minded women and, in the process, inspiring Philips's poetic creativity.

'To Amasia, holding a Burning-Glass in her Hand' by the Irish poet John Hopkins (b.1675), the younger brother of the more famous writer Charles (1671?–1700), begins in conventional tropic territory (as if Shakespeare had never written his parodic Sonnet 130):

> Whilst in your hand this Chrystal Glass I view,
> It seems almost to be as bright as you.
> Whilst your Eyes dazzling glories on it run,
> You make me fancy 'tis another Sun.

This Glass an Emblem of your coldness proves,
For that encreases, and inflames my Loves.
So, when on me your snowy hand you turn,
The solid Ice you hold, boasts Pow'r to burn.[104]

The object of desire ('Amasia') holds another object (the 'Chrystal Glass'), and their haptic connection foregrounds a form of mock-microcosmic correspondence. 'Amasia' is exposed to a kind of double figuration: she is not only the Sun (the eye of heaven which extramissively emits its beams) but also simultaneously the ice-like burning-glass, and the qualities of both inform her characterization as an inflaming yet aloof beauty. The speaker, therefore, imagines Amasia toying with his affections through her control of the glass, which becomes a proxy for her gaze. This association with tyranny then translates into a burlesque comparison, which exploits the burning-glass's mythopoeic possibilities:

Ah! charming Fair, you seem, while thus you stand,
Like Heav'n's dread thund'er arm'd, with light'nings in your hand.
Flashes from thence must vain, and useless prove,
For, who but once sees you, feels fiercer flames in Love.
The proud *Salmon'us* ne'er such light'nings threw,
As from your Silver Cloud are cast by you. (pp. 51–2; ll. 11–16)

Salmoneus, King of Elis, had reputedly tried to emulate Zeus by thundering along in his chariot on a bridge made of brass, and throwing torches to imitate lightning. Not for the last time, the burning-glass is associated with a revolutionary power that emulates and might even surpass the gods'.

As John Hall predicted, the mistress' eye/burning-glass analogy rapidly became a cliché of the wooing (or wooed) lover, recurring many times in French and English, and in poetry and visual art.[105] That the trope soon became so hackneyed can be seen in *The Fool Turn'd Critick: A Comedy* (1678) by Thomas D'Urfey (1653?–1723), where it is used by the rake 'Frank Amorous' to declare his (false) commitment to the lady Penelope: 'I am like Chaff before a burning Glass, and / Every glance from your eye converts me into flame'.[106] Here is perhaps an ironic nod to John the Baptist's warning that 'he that cometh after me … will burn up the chaff with unquenchable fire' (Mt. 3.11-12), as Amorous subtly acknowledges his deception. Once a literary conceit becomes a chat-up line, its days are numbered as a means of serious contemplation of the ways of love. John Cleland (1709–89), in *Memoirs of a Woman of Pleasure* (1749), took the simile into bawdier realms, however. His protagonist Fanny Hill is reunited with her dear Charles, whose member 'worked so strongly on [her] soul, that it sent all its sensitive spirits to that organ of bliss ... There, concentering to a point, like rays in a burning glass, they glowed, they burnt with the intensest heat'.[107] The thermo-theological discourse encountered in authors such as Horneck is adapted to allow Fanny to express the strength of feeling in her 'soul'. Cleland deploys the language of thermal optics to idealize this particular sexual union, which 'concenter[s]' both physical and mental ecstasy.

Glass vessels

In the background of much of this chapter, but brought to the forefront particularly by Horneck's union of the Sun and heart, is the slow reconceptualization of the Sun during the sixteenth and seventeenth centuries, as heliocentric gradually replaced Ptolemaic theories of the cosmos, following the publication of Copernicus's *De Revolutionibus Orbium Coelestium* (1543). In the late middle ages, the Sun was thought to be a perfect polished sphere, radiating celestial virtues throughout the cosmos, and moving in a circular orbit around the Earth. According to Aristotle, its heat was due to this motion, and was not inherent to the Sun itself. By the Restoration, the Sun had been 'desacralized', its emanations explained by atomistic philosophies, relocated to the centre, and no longer a perfect sphere but potentially an Earth-like environment with perhaps its own inhabitants.[108] While astronomers were repositioning terrestrial and solar bodies in relation to one another, artisans and natural philosophers were finding new ways of harnessing sunlight. The Copernican system, once it eventually overthrew the idea of the centre of the cosmos being the 'sump' (the lowest place), gave more impetus to the Sun's power on Earth, and the analogical connections between Sun and deity. The practical and imaginative emergence of solar technologies, therefore, formed a part of a wider refashioning of the relationship between humanity and the Sun taking place during the heliocentric era.

The connections between early modern notions of solar power and heliocentrism are most apparent in *The Comical History of the States and Empires of the Moon and Sun* (1657) by Cyrano de Bergerac (1619–55). For this fantastical, proto-science fiction voyage, Cyrano adopted many elements from Francis Godwin's *The Man in the Moone; or, A Discourse of a Voyage Thither* (1638).[109] However, Cyrano makes the Sun not only the ultimate destination of his narrator's celestial (and psycho-spiritual) journey, but also the source of energy for his flight. In so doing, he transforms contemporary ideas of solar technology to fashion his own, unique mechanisms with sublime powers that derive from but extend far beyond the ambitions of natural magic and Archimedean emulators.

Inspired by Prometheus, who 'went up to Heaven, and stole fire from thence', the engineer-narrator first devises himself a suit of 'Glasses full of Dew, … upon which the Sun so violently darted his Rays, that the Heat which attracted them … carried me up', via which he travels as far as New France, Canada.[110] In dialogue with the governor there, the narrator reveals his preference for heliocentrism, in terms reminiscent of Harvey's defence in *De Motu Cordis* (1628) (p. 10). While at this point, the narrator's personified Sun seems to be a rhetorical ploy, it later approaches a panentheistic conception of a solar deity whose radiance extends throughout the cosmos.[111] The ingenious narrator's next machine, which gets him to the Moon, involves a kind of spring packed with fireworks. Once on the populous lunar surface, however, solar energy returns to centre stage. There the narrator meets a 'Philosopher' who was 'born in the Sun' (p. 38), and possesses the skill to manipulate the energy of his former home. Their night-time conversation is illuminated by the philosopher's two transparent 'Bowls of Fire' containing 'incombustible Tapers' which are 'Rays of the Sun' purged of their heat: an early example of the fantasy of cold light encountered in subsequent centuries (see esp. Chapter 5). However, this 'Native of the Sun' is far from a natural magician, but instead a relativist, rational philosopher who wishes to dispel 'any great Cause of Admiration', because sunbeams are merely the 'Dust of that World' he comes from (pp. 118–19). Back on Earth, the narrator invents a new device:

> It was a large very light Box, … of about six Foot high, and three Foot Square. This Box had a hole in it below; and over the Cover, which had likewise a hole in it, I placed a Vessel of Christal, … in form of an Icosaedron, to the end that every Facet being convex and concave, my Boul might produce the effect of a Burning-Glass. (p. 47)

This burning-glass/barometer hybrid, reminiscent of the modern-day ramjet, is designed to unite the Sun's 'Rayes in the middle of the Vessel' and by its heat drive out the air above (p. 51). The figure of the icosahedron, a polyhedron with twenty triangular faces, was one of the shapes held in 'great estimation' ever since the ancient philosophers, grounding Cyrano's fantastical invention in Classical authority.[112] The mechanism of the 'little Chrystal-House' (p. 71) works more successfully than its inventor could have hoped for, confirming the brilliance of his geometrical design: 'In the twinkling of an Eye, the Sun which beat perpendicularly, and obliquely upon the Burning-Glasses of the Icosaedron, hoisted me up' (p. 50).

Once in space, the narrator reports that the wondrous power of the Sun is not only mechanical but also physiological and, in accordance with humoral theory, psychological: 'The more *I* advanced towards that enflamed World, the stronger *I* found my self. … and I know not what Gladness mingled with my Blood, which put me beyond my self' (pp. 51–2). This transcendence is a kind of Neoplatonic, scopic fantasy, of seeing through oneself completely, and becoming one with the cosmos through total invisibility via artificial means:

> My sight, … falling by chance upon my Breast, instead of stopping at the Surface of my Body, went quite through; … as if my Body had been no more but an Organ of sight: … my Shed and I were both become transparent. (p. 72)

At this moment of visual annihilation, the libertine narrator lifts up his eyes to the Sun, 'our common Father' (p. 74), taking a passionate belief in heliocentrism to the point of solar idolatry.[113] The foregrounding of solar technologies prior to this moment, therefore, becomes part of an extended homage to the Sun, culminating in the subject's union with the heavens' central body. Babel-like, the narrator has fashioned himself and his surroundings into a means by which he seems to touch God: perhaps the ultimate dream of the Renaissance subject, and present in many of the visions of solar technology encountered so far.

The photothermic technologies of the Renaissance fostered cosmocratic fantasies of intellectual, military or environmental domination. Control over the transformative capacity of light seemed to confirm humanity's privileged place on the great chain of being, fuelling the subjugation of nature and its resources. The rhetorical versatility of the burning-glass trope, meanwhile, gave it an ideational role, offering ways to represent and understand interior states and the construction of the self, and providing a vocabulary for the expression of faith and desire. The Sun had become the physical centre of the universe, but the reflection and refraction of its rays burnt into the heart of man.

Bundling up the Sunbeams:
Burning into the Enlightenment

Tricks of the light

One of the earliest anti-slavery fictions, *Oroonoko: Or, The Royal Slave* (1688) by Aphra Behn (1640?–89) is structured around cross-cultural encounter. This novella, possibly based in Behn's own colonial experiences, relates how the eponymous prince from the African kingdom of Coramantien is tricked into captivity and transported across the Atlantic to a plantation in Surinam. At the start of the narrative, we are given a primitivist glimpse of the 'simple Nature' of Surinamese life.[1] Once Oroonoko has been transported, both the first-person narrator (a female British colonial) and her tragic hero accompany an exploration in-land, and visit a native Surinamese village. In a reversal of the transcultural relations throughout much of the narration, the visiting Europeans become the villagers' objects of curiosity, and critics have paid attention to the way in which the eyewitness narrator describes herself as a thing of wonder.[2] Yet it is the demonstration of a burning-glass which reveals more about the relationship between the natives and the narrator:

> By the extream Ignorance and Simplicity of 'em, it were not difficult to establish any unknown or extravagant Religion among them; and to impose any Notions or Fictions upon 'em. For seeing a Kinsman of mine set some Paper a Fire, with a Burning-glass, a Trick they had never before seen, they were like to have Ador'd him for a God. (III, 102)

At this moment of apotheosis, optical technology serves as an emissary of empire, providing the narrator with material evidence of European superiority, rationality and the light of civilization.[3] This is contrasted with the villagers' ignorance and superstition, which she identifies as an exploitable characteristic for the potential colonizer, and as justification for their subjection. As elsewhere, the narrator's perspective seems split between imperialist rhetoric and sympathy for the colonial subject. Nevertheless, this passage exemplifies that the way in which Enlightenment Europe represented itself and the foreign other was bound up with its material culture. The 'Trick' of the burning-glass functions as a metonym for enlightened knowledge and yet also confers a divine identity.

This kind of transnational burning-glass encounter between the technologically advanced and less developed also became a familiar trope in non-fiction travel journals. For instance, in the 1790s, the Hudson's Bay Company trader-surveyor Peter Fidler (1769–1822) was trekking through the Rocky Mountains when he met a young Shoshone man with whom he tried to share a pipe. Fidler described how he used a burning-glass to create the spark: an apparent conjuration which terrified the superstitious young man, inciting the ridicule of the Native Americans accompanying the trading party, against their naïve countryman.[4] The burning-glass is once more aligned against the foreign other, as a marker of Western knowledge, and apparent proof of the native's inferiority. As this chapter will explore, the history of solar technologies in the long eighteenth century is punctuated by these and many other kinds of exchange and encounter, of both ideas and objects, and across nationalities, disciplines, professions and genres. Burning-glasses mediated between Paris, London and Peking, chemistry and optics, natural philosopher and artisan, factual account and fictional whimsy, as Enlightenment Europe sought to see itself in the imaginatively powerful reflections of others.

During the late seventeenth and eighteenth centuries, large burning-mirrors and lenses – made from glass or polished metal (usually steel or copper), and consisting of a single plate or series of them – consolidated their special status within the culture of natural philosophy. In a short time, these inventions moved from myth and bravado into an achievable reality, and seemed to act as material proof that the Moderns' accomplishments could equal or even surpass those of the Ancients.[5] Realizing seemingly impossible feats in their ability to transform other substances, these exalted devices were sources of intense curiosity within the republic of letters. They came out from the shadows of natural magic, and were appreciated as products of enlightened knowledge, in tandem with high-level technical skill and expertise.[6] The objects circulated in networks of natural philosophy across

Western Europe and beyond, attracting patronage from monarchs and nobles who wished to benefit from association with significant symbols of applying knowledge to the providential powers of nature. Burning-specula seemed to confirm both the plenitude of God's universe, since the phenomenal heat they harnessed demonstrated that only '*God's Almighty Hand*' could have created 'so prodigious a mass of Fire as the Sun is', *and* humanity's dominion over that glorious body and, by corollary, all of nature.[7] The glasses fascinated some of the most brilliant minds of the period, who saw their knowledge- and income-generating potential, and they highlighted how exchange across professional boundaries, between natural philosophers, inventors and instrument-makers, could achieve technological improvement. The increasing interest in, and use of, thermal optics became reflective of the expansion in the scientific instrument trade more generally. Scientific institutions were quick to gain access to and be associated with successful burning instruments, as they attempted to establish themselves as rival arbiters of natural knowledge. Crucially, the objects themselves were consumed as works of visual delight and imagination, and in literature and art they stimulated consideration of technological, epistemological and social issues.

Handy Villette and the Sun king

In the 1660s, the Royal Society's secretary Henry Oldenburg (c.1617–77) received communication of several burning-specula produced on the continent, including a mirror built by François Villette (1621–98), an engineer and fireworks manufacturer from Lyon.[8] More than 30 inches in diameter, held in a steel frame, and with a burning point 3 feet from the glass, Villette's concave could melt pot-iron in 40 seconds, and vitrify quarry-stone in 45.[9] Amid some scepticism of the device's abilities at the Royal Society, but with Robert Boyle's support, Oldenburg went ahead in publishing an account of the mirror and the pyrognostic data in the *Philosophical Transactions*, and his correspondence suggests a voracious interest in tales of rival objects that would make exciting copy.[10] Francis Smethwick's concave, for instance, could apparently burn wood in 10 seconds, but when brought to Gresham College it was quaintly tested on a glove, with an action 'mighty pretty' according to Samuel Pepys (1633–1703).[11] The rarity of the devices, however, meant that validating their extraordinary abilities relied upon eye-witnessing and reportage. These accounts also highlighted questions about a burning-mirror's ontological status (was it a legitimate object of enquiry in itself, or merely a producer of materials knowledge?), and the repeatability of the experiments.

Villette built several more concaves, and sold them to prestigious men, including Frederick III of Denmark (1648–70). The Académie Royale des Sciences saw the utilitarian possibilities of Villette's inventions, and his largest, of 34 inches diameter, became subject to numerous trials.[12] In the summer of 1669, Louis XIV (1638–1715) watched the concave perform 'with great satisfaction' and purchased it for his own collections and the use of the Académie Royale.[13] On display in the Marais district, urban residence of the aristocracy and nobility, the mirror became a technological addition to the absolutist iconography of the Sun king, amid the other pyrotechnical and solar symbolic displays of the Ancien Regime. The Louvre, the Tuileries, Versailles and the Paris Observatory all contained areas designed in accordance with solar symbolism. For instance, the *Grotte de Thétis* at Versailles, built in 1664, had been designed by brothers Charles and Claude Perrault (1628–1703, 1613–88, respectively) to integrate the movement of the Sun, with its light manipulated by grills and mirrors.[14] Villette's mirror, however, went one stage further, manifesting the king's power both visually and physically. In Sébastien Leclerc's engraving of an imaginary visit to the *Bibliothèque du roi* of the Académie Royale by Louis XIV, Villette's mirror is a central spectacle, with the king standing directly before it, flanked by his brother the duc d'Orléans (1674–1723) to his right, and the treasury minister Jean-Baptiste Colbert (1619–83) to his left.[15] The illustration by Leclerc (1637–1714) attempted to legitimate this emerging society via association with the king's glory, and Villette's mirror intertwines reflections of cosmic, monarchical and institutional power.

Conventional, decorative mirrors, both beautiful and expensive, had become symbols of aristocratic opulence, and Louis's court was obsessed with their splendour, and the opportunities for social display and self-inspection they triggered. Colbert had founded the Manufacture royale de glaces de miroirs in 1665, in an attempt to emulate Venetian manufacturing quality, and this investment culminated in the Hall of Mirrors at Versailles, where the court became a spectacle unto itself. Villette's burning-mirror offered a different specular experience: not one of clarity, reciprocity and self-admiration, but otherworldly power, visual distortion and destruction. Rather than creating pure images of reflection, giving one knowledge of the self, Villette's mirror instead distorted the gaze, reversing the coenaesthesia (the awareness of one's bodily state) associated with what Sabine Melchior-Bonnet calls the 'rise of specular consciousness'.[16] Crucially, these images involved a literal 'immersion' of the spectating subject outside who looks into, but is also seemingly materialized within, the 'aesthetic terrain' of the mirror.[17] According to the visiting British diplomat Francis Vernon (bap.1637, d.1677), 'People flock to see it with Incredible Curiositie' as 'it is extreme strange to see ... with what vigour the Rayes penetrate & dissolve' stone or slate. While this photothermic ability seemed to materialize the sunbeams and suggest utilitarian possibilities, the concave's phantasmic effects appeared to be as much, if not more, of an attraction. For instance, during the 'riskless duel' trick, adapted from della Porta, 'a staffe ... wch one sees on the superficies of the glass shall seeme to bee thrust out, ... soe that It makes most start & expresse suddain alterations of feare as if they were toucht wth it'.[18] The multiple effects of reflection created sublime and hallucinatory spectral images that added to the sense that the mirror's powers were unearthly. The king himself, however, was not amused by the 'riskless duel' when requested to draw his own sword, damaging the representational effects the mirror was purchased to promote: 'his image appeared to leap forward and direct a thrust at his own face, the great monarch recoiled in alarm' and was 'so much ashamed of himself directly afterwards' that he had the mirror taken away.[19] Notwithstanding this incident, the place of Villette's mirror within Louis XIV's court, where it was celebrated for its protean (and not necessarily 'scientific') abilities, confirms that the aesthetic contours of knowledge-making are often essential to its circulation and, through patronage, sustaining its production.

Villette's portrait is a fantasy of haptic control, reducing down the mirrors he produced in order to depict him Prometheus- or God-like, with the power of the Sun in his hand (Figure 5). The image of God's, and subsequently the monarch's, powerful, often healing hand dates back to Classical times, and scientific illustrations during this period sometimes conferred a similar kind of authority upon the hand of the heroic philosopher.[20] However, the verse inscription tells us there is perhaps a more specific, less mythical reason for this particular diminution: 'From the hand of famous Villette / Everybody sought after / Burning mirrors magnifying glasses' and that Villette was an 'upright man and without vice / Skilful without being arrogant'. As we will

Figure 5 Engraving by Etienne Jehandier Desrochers, frontispiece to *Description du Grand Miroir Ardent, fait par les Sieurs Villette Pere et Fils Natifs de Lion* (Liège, 1715). Wellcome Collection CC BY.

see repeatedly, to aspire to possess the Sun's power is often represented as an egotistical, tyrannical Promethean ambition, yet the verse here celebrates Villette's heroic handiwork and humility, manual dexterity and morality. There is little evidence that Villette was himself involved in the moulding and polishing of his devices in the way that, say, William Herschel did with his telescope mirrors a century later. This individual apotheosis, therefore, perhaps obscures the collective labour behind the mirrors; nevertheless, this representation seeks to valorize the application of natural philosophy.

Making light matter

Advances in thermal optics continued on the continent, buttressed by ancient myth, royal patronage and popular fascination. Meanwhile, the Royal Society continued to show a somewhat sporadic interest.[21] In 1687, the Society learnt of a copper-plate concave made by an anonymous German inventor and capable of generating such extraordinary heat that 'a piece of Tinn or Lead three Inches thick, as soon as it is put into the *Focus*, melts away in drops'. Robert Hooke (1635–1703) wanted the Society to construct something on a grander scale, although there is no evidence this suggestion was acted upon.[22] The German inventor, Ehrenfried Walter von Tschirnhaus (1651–1708), Lord of Kislingswald and Stoltsenberg, was a mathematician, philosopher and friend of Spinoza and Leibniz, and his work ultimately resulted in the development of Meissen porcelain for Augustus II, Elector of Saxony and King of Poland (1670–1733). Leibniz (1646–1716) and Spinoza (1632–77) had recommended Tschirnhaus to Colbert as a tutor for his son, and while in Paris in the late 1670s he was seemingly enraptured by Villette's mirror. On a trip to southern France and Italy, Tschirnhaus visited various glass manufacturers, including the burning-mirror designers Villette (in Lyon) and Manfredo Settala (1600–80), a cleric and virtuoso in Milan, whose mirror was reportedly 7 feet in diameter and could burn at a distance of 33 feet. By 7 April 1681, Tschirnhaus was writing to Leibniz about his method of making large mirrors and, significantly, his theory of caustics.[23] Tschirnhaus discovered that 'Causticks are Curves, formed by the Concourse of the Rays of Light, which … are equal to known Right Lines, when the Curves that produce them are Geometrical'.[24] Despite Tschirnhaus's mathematical innovation, it was the application of the burning-glass which became his main interest. Just as the Royal Society was learning of his mirrors in the late 1680s, Tschirnhaus foresaw that 'larger and better Convex Glasses exposed to the Sun, would be a kind of new Furnaces, and consequently produce a new Chymistry'. To achieve this, however, required not only the skill to polish copper plates but also an oven which could be accurately controlled, to cool down the glass slowly over a period of weeks. Tschirnhaus convinced Augustus II to provide him with a glass-house in Dresden, and he began to work with Johannes Kunckel (1630–1703), a glass-making expert then serving as court alchemist.[25] In November 1691, Tschirnhaus announced his new burning-lens, an invention which apparently 'puzzle[d] the most skilful People, [as] nothing is a more noble Panegyrick upon Monsieur Tschirnhaus's Skill in the Mechanicks than this'.[26] Tschirnhaus put his lens to work in investigating porcelain, managing to melt it in early 1694. Yet it would not be until fourteen years later (9 October 1708), two days before his death, that his factory fired its first cup of true unglazed porcelain.[27] The surviving portrait of Tschirnhaus, dated to the year of his passing, like Villette's, is a celebration of his technological mastery. It places one of Tschirnhaus's mirrors centre stage, capturing a mirror image of the man himself, who is seated to the left-hand side, turned away from the painter. The portrait invites us to view it using a mirror oneself, in order to see the reflection of the reflection, and thereby the original, un-inverted genius of Tschirnhaus.[28]

In 1703, the Royal Society was informed of another burning-glass ('verre Brulant') built by Tschirnhaus, which had been bought by Philippe III, Duc d'Orléans (1674–1723) and tested by Etienne François Geoffroy (1672–1731).[29] Comprising two lenses, the first of 3 feet diameter with the refracted rays then contracted in a second lens of an inch diameter, the device could apparently reduce a piece of gold to a 'transparent Sky coloured Stone'.[30] For Geoffroy, the remarkable instrument became an important piece of apparatus because it 'far surpasses the force of our common Fires', allowing unprecedented analysis of the sulphur content of metals undergoing vitrification, and crucial in facilitating other research towards his tables of chemical affinity. Geoffroy admitted, however, that he had been hindered by poor weather, and on the days it was working, he had difficulty finding crucibles which could withstand the intensity of the heat without disintegrating themselves.[31]

Tschirnhaus's latest invention encouraged Isaac Newton (1642–1727) (by then president of the Royal Society) to disclose at a meeting of the Royal Society 'that he had thought of a Contrivance for Burning Glasses by uniting severall'.[32] Newton's design comprised six foiled glasses surrounding a seventh, each twelve inches in diameter, 'so plac'd about the middle Concave, that they prict their illuminated points all at once'.[33] The set of glasses was promptly constructed and tested in the middle of 1704.[34] The *Opticks* was published this same year, and its theories faced substantial opposition from continental philosophers.[35] British Newtonians, such as John Evelyn (1620–1706) and William Derham (1657–1735), put the burning device forward as evidence of the successful practical application, and hence legitimacy, of Newton's ideas, and argued it was more powerful and cheaper to construct than its European rivals.[36] The summer of 1705 was 'exceedingly dry & hot' (Evelyn, 17 June 1705; *Diary*, V, 599), allowing Newton to test his device extensively. According to Newton's friend David Gregory (1659–1708), Savilian Professor of Astronomy at Oxford, gold when placed in the focus '[boils] up into Bubbles, [then] flyes out into litle round Gold spherules'. Newton apparently intended for twelve other glasses to encircle the current seven, thinking this would double the intensity. There is no evidence this modification was carried out, but the prospect allowed John Harris (c.1666–1719) to speculate 'If another round of Concaves were added about these seven, perhaps it would out do any thing we have any Account of'.[37] For reasons unknown, the speculum disappeared from the Society's Repository sometime between the 1730s and 1760s.[38] The intellectual legacy of Newton's burning-glass is also vague. Designing and experimenting with his own device (or reading about the lenses and mirrors of others) may have influenced Newton's theories of radiant heat, added to the 'Queries' in later editions of the *Opticks*, but there is no conclusive proof, and Newton himself remained tight-lipped about his invention, other than to Society members.[39] Tschirnhaus's interest in geometrical curves, however, has been linked to his practical work in catoptrics.[40]

While burning-mirrors and lenses were sometimes claimed, in the support of philosophical and/or national allegiances, to be the result of theoretical knowledge rather than hands-on expertise, the objects themselves became important to the advancement of corpuscular hypotheses. In his 'New Theory about Light and Colors' (1672), Newton tentatively suggested that because colours are qualities of light, the rays themselves cannot also be qualities, but must be substances.[41] Over the next half-century, apparent in changes to the 'Queries' in the *Opticks*, Newton gradually came to accept an emission theory in which 'Rays of Light' consisted of 'very small Bodies emitted from shining Substances', and moving at the immense velocity discovered by Ole Römer (1644–1710).[42] Other natural philosophers were less cautious, however. William Molyneux (1656–98), founding member of the Dublin Philosophical Society, argued that because light shared properties with matter then it must be a body itself, a statement Harris repeated in his popular *Lexicon Technicum* (1704). For Molyneux, light's corpuscularity was especially evident in its behaviour in a burning-glass. The rays are 'crowded close' in the focus, with the glass 'bringing the Light, that naturally should have been diffused through some other part, to the more enlightened place'.[43]

At around the same time, Wilhelm Homberg (1652–1715), who had initially attempted to employ the duc d'Orléans's Tschirnhaus glass as a way of transmuting silver into gold, was using it to investigate the 'sulphur principle' in fire. Calcinating various substances with heat from the lens, Homberg attributed the weight gained by some materials when they were burnt to their interaction with the 'Matter of Light'.[44] Separate experiments involving the 'Bologna Stone' (phosphorescent barite) and the calcination of antimony using the burning-lens convinced Homberg that sulphur was light which had been incorporated into matter.[45] This was celebrated by the Paris Academy as unlocking '*Physicks almost New ... into the internal Composition of Bodies*', and a 'new Chymistry, the only Furnace of which, is the Burning-Glass in the Royal Palace'.[46] Human touch of the intense beams focused by burning-glasses also suggested the validity of mechanical accounts of light's transmission: the lens manufacturer William Parker described the sensation as being like 'a *sharp cut* with a lancet'.[47] As a mediating device, the burning-glass seemed able to concretize the elusive and ephemeral sunbeam, producing a kind of 'reality effect'.[48] If the Sun's rays could be focused by an instrument and put to work upon matter, the concept of light as a material projectile could be understood as a real proposition rather than a rhetorical ruse, transforming metaphor into identity.[49]

Meanwhile, as we saw in the previous chapter, the humble handheld burning-glass had become a fashionable accessory within polite culture, and a prolific generator of metaphor. As a symbol of intensity, focus, mediation, precision and penetration, this gadget became a useful trope in relation to themes of burning desire or religious devotion, uniting fragmentary feelings into a single passion or bringing light to the dark interiority of the soul. The imaginative value of this device extended beyond the tropic, however, contributing to debates about man's place in the cosmos, mechanical theories of mind and technological progress.[50] In these moments, the handheld glass transcended the quotidian, to become a sublime technological spectacle in itself.

Meditation, myth and celestial satire

In the eighteenth-century literary imagination, representations of solar devices exploited the relational meanings between the large and handheld mirrors, their mythopoeic potential and their ability to move across cosmological scales. A significant example can be found in *The Tatler* No. 100 (29 November 1709), attributed to Joseph Addison (1672–1719), essayist and doyen of the London literary circle. This piece predates but begins in the manner of Addison's *Spectator* essays on the pleasures of the imagination and natural theology, with the narrator recalling how he beheld the night sky 'so wonderfully adorned and lighted up' that it prompted 'Meditations' on the Creator. In a slumber at his lodgings he then had a 'Vision', Classical rather than Christian in nature, in which an interplanetary woman resembling Astraea, goddess of truth and justice, descends to earth wielding a mirror with magical incendiary powers, 'endowed with the same Qualities as that which the Painters put into the Hand of *Truth*':

> The World was in an Alarm, and all the Inhabitants of it gathered together upon a spacious Plain; ... A Voice was heard ... declaring the Intention of this Visit, which was, to restore and appropriate to every one living what was his Due.[51]

As one would expect from *The Tatler*, this dream of mythological apocalypse becomes an occasion for satire:

> The Inhabitants held up the Instruments of their Tenure, whether in Parchment, Paper, Wax, ... and as the Goddess moved the Mirror of Truth ... The Rays [set] Fire to all Forgery and Falshood [and] very often ran through two or three Lines only, and then stopped. ... the Light ... pierced into all the dark Corners and Recesses of the Universe, ... This occasioned a wonderful Revolution among the People. ... the Spoils of Extortion, Fraud and Robbery ... were thrown together into a prodigious Pile, ... to which all injured Persons were invited to receive what belonged to them. ... What moved my Concern most, was, to see a certain Street of the greatest Credit in *Europe* from one End to the other become Bankrupt. (*The Tatler*, II, 116)

The mirror's panopticon-like gaze penetrates the limits of the universe, reminiscent of Epicurus 'piercing' to expose nature's secrets in Lucretius's *De Rerum Natura*.[52] Far from merely bringing things to light, in the hands of the goddess, the mirror is a precise and discriminating instrument, 'surgically removing' the false lines, exploiting the vulnerability of paper and ink as the vehicle of law, in order to restore truth and justice.[53] The narrator then expresses ironically his concern about the Bankers of Lombard Street, whose wealth has been built upon a deluge of fraudulent transactions.

The mirror's lustre is also too bright for those who have not distinguished themselves in politics, knowledge, writing and business, and so 'posts of dignity and honour' are conferred on persons of merit. It also inspires children to find their natural fathers. Like Villette's mirror, Astraea's glass creates more than photothermic effects, and in the continuation of the story in No. 102 (3 December 1709), the 'Celestial Light of the Mirror ... banishes all false Appearances, and show[s] People what they are' (*The Tatler*, II, 126). Drawing upon the medieval magical tradition, and distorting appearances as Villette's devices did to the Parisians, the ugly forms viewed in goddess's mirror provide an index to moral depravity.

Optics was (and remains) particularly inspiring for satirists, providing a physical manifestation of the different perspectives and magnifications satire offers. Moreover, the mirror had long been adopted as a metaphor for the satiric mode's ability to expose, with a view to correcting, the errors of the self.[54] Addison's burning-mirror, transformed into an instrument of apocalypse, is literally a brilliant device for fantasizing about the agency of satire: that this mode has the potential to not only expose but also eradicate error and corruption on a universal scale, just by focusing its gaze in the right direction. And yet, tellingly, it is all just a dream: Addison concedes that satire's powers could not be so revolutionary.

Issuing Astraea with a burning-glass would be repeated in Francesco Cepparuli's illustration 'Truth Opens the Eyes of the Blind' (1744).[55] Mixing Classical with Christian motifs, the engraving is accompanied by the citation of Ps. 34.5 ('They looked unto him, and were lightened: and their faces were not ashamed'), and depicts the goddess of truth using a handheld convex lens to focus sunlight upon the multitude: the righteous bask in the light, while the wicked hold their heads in shame, or even tumble into darkness on the ground. Throughout *The Tatler*'s story, meanwhile, the Sun itself plays no part in the production of the rays: they originate from within the mirror itself, making it one of the first 'pseudo-solar' technologies in the cultural imaginary. The story is also intriguing in its treatment of gender. As we have seen so far, the real mirrors and lenses had all been invented and operated by male natural philosophers and instrument-makers. Alongside them are found poetic and visual images of the beautiful woman wielding the glass as an extension of their attraction as a sexual object, enflaming male desire.[56] In contrast, displaced into the world of myth, the mirror in Addison's story becomes part of a feminized technological sublimity bringing socio-economic revolution, subverting the reactionary trope. Nevertheless, the burning action of Astraea's mirror (and its function as a revealer of hidden truths) is clearly inspired by the devices circulating in Europe at this time. This mirror is also a figure of revelation, renewal, liberation and the restoration of knowledge: connotations that surrounded the real devices too, especially as the eighteenth century progresses, and scientists such as Joseph Priestley (1733–1804) and

Antoine-Laurent Lavoisier (1743–94) use them to make important discoveries, particularly in pneumatic chemistry.

Like Addison, the greatest satirists of the period – Jonathan Swift (1667–1745), Alexander Pope (1688–1744) and their 'Scriblerian' circle – also found much satiric and comic potential within the burning-glass conceit, including ways to laugh at themselves. For example, according to Pope, in the summer of 1714 Swift discovered a new 'pastime':

> He has in his window an Orbicular Glass, wch by contraction of *the* Solar Beams into a proper Focus, doth burn, singe or speckle white or printed Paper, in curious little Holes, or various figures. We chanced to find some Experiments of this nature, upon the votes of the House of Commons. ... I doubt not but these marks of his are mysticall, *and* that the Figures he makes this way are a significant Cypher.[57]

Pope's 'very comical accompt', sent to their mutual friend Dr John Arbuthnot (1667–1735), the mathematician, physician in ordinary to Queen Anne, and fellow of the Royal Society, presents Swift's 'Experiments' as pyrotechnical parodies of Newton's optical trials.[58] The Scriblerians were immersed deeply in the political machinations of the Tory government at this time: Swift had spent 'four years jading himself with Ministers of State'.[59] The joke that Swift has created mystical 'Figures' thermographically draws upon the currency of fears about Jacobite rebellion and the ministerial rivalry between Robert Harley, Earl of Oxford (1661–1724) and Henry St John, Viscount Bolingbroke (1678–1751). Moreover, there are 'shared metaphorical entailments' between the burning-lens and other forms of optics at work in Pope's joke, although the metaphor is taken one stage further: Swift's inflammatory satiric gaze is externalized through the 'Orbicular Glass' and its destructiveness is literalized on paper.[60]

Arbuthnot subsequently divulged that he had been inspired to 'make an EpiGramm' about Swift destroying the manuscript of his controversial *History of the Four Last Years of the Queen* 'wt a Burning glass'. Mythologizing Swift's writerly act, the god of the Sun and of poetry, Apollo proclaims that 'perceaving that a faction who could not bear their deeds to be brought to light had condemnd it to an ignominious flame, that it might not perish so, he was resolv'd to consume it wt his own a celestial one'.[61] The burning-glass is again associated with the potential to not only bring hidden things into the realm of knowledge but also erase and punish. Arbuthnot's epigram, if ever written in earnest, is unrecovered, but several related works survive. *A Letter to a Young Poet* (1721), probably by Swift, compares the burning-glass's ability to the 'easie Method … of Abstracts, Abridgments, and Summaries, &c. which … collect the diffus'd Rays of Wit and Learning' ('index learning' being a frequent Scriblerian bugbear), while the *Memoirs of Martinus Scriblerus* mentions the protagonist's 'Method to apply the Force arising from the immense Velocity of *Light* to mechanical purposes' in a list of his many audacious schemes.[62] The political connotations of celestial flame also sprang to mind a few years later when Swift, back in Trim (Ireland), watched a particularly vivid display of the aurora borealis:

> The ray which shot flew like lightening and flasht all over the Sky, and darted as we agreed, like the Rays from a Looking glass when you turn it against the sun, as Boys do out of a Window … This Appearance is for Sueden as M[rs] Peggy says, as that of last year was for the Pretend[r].[63]

The phenomenon receives a mixture of responses: Swift's use of technological analogy as an attempt to explain, and the superstitious housekeeper Peggy Dixon's omen-production, since Sweden had recently emerged as a threat against Hanover and Britain.

The myriad of connections between the construction of burning-glasses, corpuscular theories of light, and the cultures of invention, experiment and projecting are played out more fully in *The Humble Petition of the Colliers* (1716), the technological satire discussed in the opening of this book. Published as an anonymous single folio sheet, this parody is assumed to have been written at least in part by Arbuthnot. Indeed, the Scriblerians were acutely aware that their learned friend possessed a unique ability to turn the latest developments in natural knowledge into conceits with satiric power.[64] Arbuthnot probably had first-hand awareness of large burning-mirrors: he may have witnessed one of Villette's concaves when visiting Paris in 1699, and was a member of David Gregory's circle which had discussed Newton's device intensely.[65] In addition, *An Essay on the Usefulness of Mathematical Learning* (1701), traditionally ascribed to Arbuthnot, celebrated the variety of 'useful inventions' furnished by catoptrics and dioptrics, including lenses designed 'to produce heat unimitable by our hottest furnaces'.[66]

The protectionist Petitioner argues that the 'certain *Glasses*' which 'bundl[e] up the Sun-Beams' would reduce cooks and blacksmiths 'to Beggary' and force colliers out of business.[67] Displacing traditional occupations and artisanal knowledge, the proposal would 'throw the whole Art of *Cookery* into the

Hands of Astronomers and Glass-Grinders', while cooks themselves would be obliged to 'study Opticks and Astronomy'. Reminding readers of the total solar eclipse in April 1715, and resembling other Scriblerian satires involving mock-apocalyptic destruction, infection or metamorphosis, the Petitioner fears what might happen if an 'Eclipse of an Extraordinary Length' deprived London of the 'Sun-Beams for several Months' (p. 2).[68]

Within *The Humble Petition* there is an overt resistance to any description of the burning-mirror. Yet this negation is probably a symptom of genre. The English patent system at this time was principally concerned with the registration, rather than the examination of inventions, and so most applications kept their subjects in obscurity to prevent possible competition.[69] As a parody of published objections to the soliciting of patents, the work impersonates a petitioning tradesman kept ignorant of what these 'certain Glasses' actually involve, but who is anxiously concerned about their possible effects. So, the Petitioner follows his own warped logic when outlining the risk to public health, the church and state from the poor nutritional value and toxic effects of 'catoptrical cookery'. Caricaturing celestial theories of disease, ingested sunbeams would produce 'great Quantities' of 'Inward Light', encouraging Quakerism and poetry, or mean that 'the influences of the Constellations ... will ... be convey'd into the Blood; and when the Sun is amongst the Horned Signs, may produce ... a Spirit of Unchastity' (p. 2).[70] The Petitioner's fears that the 'catoptrical victuallers' will 'Monopolize the Beams of the Sun' (p. 1), and that 'Seeds and other parts of Plants' will be 'impregnated with the Sun-Beams' (p. 2), burlesque contemporary theories about the finite quantity of the Sun's light, and corpuscular concepts of light more generally. Arbuthnot's acquaintance George Cheyne (1671/2–1743), for instance, had popularized John Bernoulli's discovery that 'all the Bodies on our Globe are *saturated* at all times with Rays of Light which never return again to their Fountain, because ... Bodies do attract, and consequently retain these Rays'.[71] It is unlikely that *The Humble Petition* is satirizing such ideas, however, but distorting them parodically in order to revel in the Petitioner's naïvety.

The Petitioner's absurd anxieties are diagnosed as originating in ignorance of technology coupled with misunderstood conceptions of nature, but which are seemingly generated by the difficulties of representing natural knowledge. Nevertheless, this impersonation gives licence to expressing legitimate concerns about the economics and ethics of new technologies, producing a double edge that has generated critical disagreement over what exactly this 'technophobic' parody intends to satirize: commercial projectors, reactionary proto-Luddite groups of workers, or both.[72] Significantly, *The Humble Petition* (1716) had appeared when the Longitude Act (1714) was seeking to reward financially the development of practical solutions to determine longitude at sea, part of a broader growth in practical and technological projects claiming to apply natural knowledge. Some contemporaries worried, however, that many of these schemes were poorly designed or even fraudulent, and the Scriblerians drew upon anxieties about corrupt, mercenary projectors in their satires.[73] Indeed, *The Humble Petition* associates the catoptrical cookery venture with the impractical, and often ridiculed, rocket-based global-positioning scheme of William Whiston (1667–1752) and Humphrey Ditton (1675–1714) (p. 2). Moreover, the dishonesty (or at least self-deception) of the 'CATOPTRICAL VICTUALLERS' is made apparent in *The Humble Petition* several times. For instance, they propose to use 'the *Moon* by Night, as of the *Sun* by Day' (p. 1), while Arbuthnot was no doubt aware that many natural philosophers at this time had tested the 'doctrine … that there is no Light without Heat' by exposing their burning-glasses to the full moon, but found no effects.[74] For the naïve Petitioner, the use of light produced by lunar reflection is an object of real fear, as it will 'utterly ruin the numerous Body of Tallow-Chandlers'.

Whatever the precise trajectory of its ironies might be, *The Humble Petition* crystallizes cultural and social questions bound up with the 'real' burning-glass projects themselves, concerning the commercialization and dissemination of natural knowledge and technological innovation, the adoption of mechanical labour, and the proximity of technical and artisanal expertise. Arbuthnot's jeu d'esprit serves as compelling evidence for the unique contribution that satire can make, as a kind of cultural arbiter, within the social circulation of natural knowledge.

If Swift did not himself make a contribution to *The Humble Petition*, then it is very likely he was inspired by it when composing Lemuel Gulliver's visit to the 'Academy of Lagado', a research facility hosting a number of absurd and impractical projects. The first man he meets, with a long, ragged beard 'singed in several places', is like a 'catoptrical victualler' who has seen better days:

> He had been Eight Years upon a Project for extracting Sun-Beams out of Cucumbers, which were to be put into Vials hermetically sealed, and let out to

warm the Air in raw inclement Summers. He ... did not doubt in Eight Years more, that he should be able to supply the Governors Gardens with Sunshine ... but ... intreated me to give him something as an Encouragement to Ingenuity, especially since this had been a very dear Season for Cucumbers.[75]

He is not the kind of exploitative profiteer Swift's contemporaries anxiously railed against, but a desperate, deluded experimenter.[76] Nevertheless, the man adopts a projector's patter intended to secure investment: firmly predicting his future success within a specific timeline, and appealing to Gulliver's goodwill to reward his 'ingenuity' so far, despite no evidence of a breakthrough, and with no potential market for this freely available 'commodity'.[77] It has been suggested that Swift's idea of extracting sunbeams from cucumbers twists two potential sources: work on plant respiration as printed in the *Philosophical Transactions* and Thomas Shadwell's satirical play *The Virtuoso* (1678), in which Sir Nicholas Gimcrack seeks to sell bottled country air to gullible Londoners.[78] As we have already seen, *The Humble Petition* might have furnished Swift with his particular conceit, and popular physico-theological books such as Cheyne's were speculating that 'all the virtual Heat in the juices of Vegetables, Metals and Minerals, may be owing to the imprisoned Rays in 'em' (*Philosophical Principles of Natural Religion*, p. 95). Like Arbuthnot before him, Swift exploits the reductiveness of corpuscular theories of light in order to interrogate motives behind the commercial application of natural knowledge, each connected to a potentially suspect materiality.

Demonstration, delight and discovery

During the 1710s, Villette's mirrors not only continued to astonish and entertain but also encountered suspicion and superstition, making the humble Petitioner's concerns not so far-fetched. Étienne-Gaspard Robert (1763–1837) recounts how Villette arrived in Liège in August 1713 at a time of intense rain that spoilt the year's harvest and forced up the price of bread. A frenzied mob, convinced that Villette's mirror had caused the bad weather, sought to advance on his residence to break his invention. Joseph Clement, Archbishop of Cologne (1671–1723) intervened, arguing that only God possessed the power to open and close the heavens.[79]

In 1718, British Newtonians got the chance to test one of Villette's concaves when it was brought to England by one of his sons, and displayed in the Privy Garden at Whitehall. This mirror was almost 47 inches in diameter, and made from a specially composed compound metal.[80] The Privy Garden, already the home of numerous sundials (including Charles II's), was a site that opened out the mirror to arguably a wider social audience than ever before: at half-a-crown a person, it could be viewed daily from seven in the morning till six at night.[81] Moreover, the activities conducted with the mirror, observable by the paying visitors, were not merely 'popular' rational amusements but also attempted to verify the reactions described in Villette's account. John Harris (whose *Lexicon Technicum* had venerated Villette's work, but praised Newton's burning apparatus above all others) and John Theophilis Desaguliers (1683–1744), the Royal Society's Curator of Experiments, tested standard substances such as tin (which melted almost immediately) and unique materials from antiquity (such as 'A piece of *Pompey*'s Pillar at *Alexandria*', which vitrified in around fifty seconds).[82] Spectators also enjoyed how the device could displace and distort images so to act like a camera obscura or hall of mirrors, representing 'all Things so naturally, that the Field or Garden seems to be within the Room', or giving each viewing subject two chins and three eyes.[83]

The potential of the burning-mirror as a dedicated entertainment device to a public becoming increasingly interested in mechanical and natural wonders was realized by George Willdey (bap.1676, d.1737), Master of the Spectaclemakers' Company, whose 'Great Toy, Spectacle and Print Shop; or, Grand Magazine of Curiosities' was on the corner of Ludgate Street, next to St Paul's. Willdey has been called the 'most prolific advertiser for items of public science', filling the newspapers week after week during the 1710s and 1720s.[84] Seeking to capture the interest in Villette's mirror, Willdey announced he had 'finished the best Burning Glass in the World, and plac'd it upon the Top of my House'. After making the usual sort of claims about its ability to vitrify various substances, he also suggests its application as a 'Sun Kitchen, where meat may be Boil'd, Bak'd, Roasted, Stewed or Broil'd. Coffee, Tea, Chocolate made'. (*The Humble Petition*'s 'catoptrical cookery' made real.) Willdey then directly addressed its competition, claiming that his glass 'far exceeds that show'd in the Privy Garden in White Hall, though each Person paid Half a Crown', while his own attraction was free to those who spent at least 5 shillings in the shop beforehand.[85] The 'great Burning Glass' Samuel Johnson (1709–

84) enjoyed visiting many years later was probably Willdey's, suggesting the continued success of its marketing.[86] On both sides of the channel, therefore, the burning-glass was assimilated into pleasure garden culture, where it oscillated between philosophical demonstration and popular entertainment, endowing viewers with knowledge and delight. As the chapter later explores, the burning-glass both anticipated and directly inspired the phantasmagorical technologies that emerged later in the century.

As we have witnessed so far, the large burning-mirrors and lenses that brought reverie and discovery across Europe were of uncertain status within the culture of natural knowledge. At mid-century, the heroic feats of Archimedes's mirror still prevailed as a hallowed, if unproven, moment of technological mastery in Enlightenment circles, occupying the minds of several theoreticians and philologists; while visitors to Syracuse would be shown the tower upon which the mirror reputedly stood.[87] In 1747, when endeavouring to construct an instrument to measure the intensity of light, Georges-Louis Leclerc, Comte de Buffon (1707–88) designed a 'sort of *Polyhedron*' comprised of around 150 small plane mirrors, held in a wooden frame of 6 feet square. Each mirror was controlled by 'three moveable Screws, … so contrived, that the Mirror can be inclined to any Angle' in order to unite with the others, although in practice this was difficult to achieve.[88] Tested at the Jardin du Roi, Buffon's invention (similar to, but apparently not inspired by Kircher's device) was by some believed to have re-established the credibility of a story thought to be 'impossible and romantic', simultaneously an ancient artefact and a novelty.[89] It is '*Archimedes* revived', noted one observer.[90] Like Kircher's work, however, Buffon's designs also proved an inspiration to others, including Étienne-Gaspard Robert.

Emerging from within the tradition of natural magic, innovations in thermal optics, therefore, continued to be valued as much for their contribution to spectacle, and confirmation of myth, as their use within experiment. The same case has been made for many seventeenth-century instruments,[91] but the primal symbolism of harnessing the Sun's power fused acutely the binaries of myth and modernity, incorporeal and material, divine and human, artifice and nature, and thus conferred a unique cultural identity upon the burning-glass. The device's liminal quality is visualized in Jacques de Lajoüe's (1687–1761) painting 'L'Optique' (1737?) (Figure 6). Commissioned by Ferdinand d'Albert d'Ailly, duc de Picquigny, the painting locates modern optical instruments

Figure 6 Charles Nicolas Cochin, line engraving of 'L'Optique. Tiré du cabinet de Monsigneur le Duc de Picquigny', after a painting by Jacques de Lajouë (Paris, (1737)). Wellcome Collection CC BY.

within a classical setting. At its centre stands a burning-mirror reflecting an intense shaft of light that leaves its human observers cowering in the shade, accentuating the burning-mirror as an emblem of enlightened knowledge and yet the mediator of a transcendental power beyond rationalization. While Edmund Burke (1729/30–97) found darkness 'more productive of sublime ideas', he conceded that 'Extreme light, by overcoming the organs of sight, obliterates all objects, so as in its effect exactly to resemble darkness'.[92] The burning-glass, therefore, reified the sublime destructive power of light.

During the eighteenth century, burning-glasses proved unquestionably their usefulness as knowledge-making tools, since they were responsible for 'the greatest degree of heat producible by man'.[93] They became part of the appareil of Enlightenment chemistry: favoured over charcoal furnaces as laboratory heat sources because they could reach higher temperatures, did not produce their own fumes and, when used in conjunction with suitable apparatus, allowed substances to be heated in glass *in vacuo* and the vapours collected. Hooke, and then some years later, Francis Hauksbee (the elder) (bap.1660, d.1713) and Jean Bernoulli (1667–1748) had fired gunpowder using them, for instance. Hooke had also found the light of the burning-glass useful when examining specimens in the microscope, effectively creating a precursor to the solar microscope considered to have been invented by Nathaniel Lieberkuhn (1711–56) in the 1730s, which projected magnified images to the delight of spectators.[94] For Priestley and Lavoisier, though, the glasses were crucial to their most important investigations. Indeed, on one level the issue of whether bodies contained an element called phlogiston, which was lost during combustion and absorbed by the air, concerned not just chemical identification but also the mastery of instrumentation.

In the summer of 1772, Lavoisier and several other academicians requested for the large Tschirnhaus burning-lens that had been bought by the duc d'Orléans to be taken out of storage so that they could investigate the destruction of diamond and other materials at the highest temperatures possible. This 'lentille du Palais Royal' was used in series with a second Tschirnhaus lens (of the same diameter but shorter focal length), producing an awesome machine (Figure 7).[95] Lavoisier's collaborators included Pierre Joseph Macquer (1718–84), who

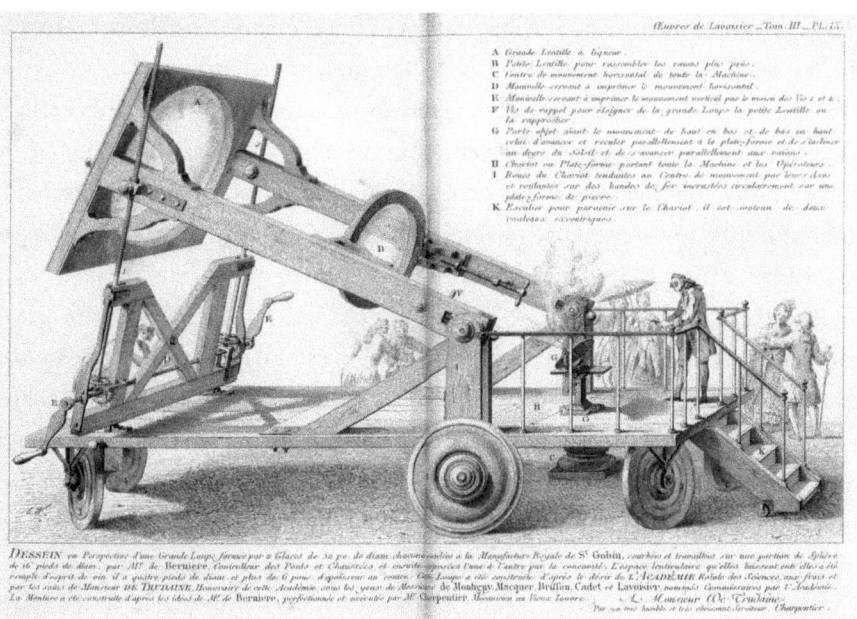

Figure 7 *Oeuvres de Lavoisier*, ed. Édouard Grimaux, 6 vols. (Paris: Imprimerie Imperiale, 1862–93), III, Plate 9. Wellcome Collection CC BY.

from the late 1750s had been attempting to melt platinum via concentrated solar rays.[96] As part of his work, Lavoisier read the papers of Geoffroy and noticed that his predecessor had not attempted to capture and test the vapour emitted by metals when heated by the burning-lens. Challenging the view that a metal consisted of a 'vitrifiable earth combined with an oil or an inflammable principle', Lavoisier conjectured that metals must contain air and that this was released during calcinations. However, he was unable to procure glass vessels that could withstand the heat of the lens.[97]

Meanwhile, Priestley was tracing the work of Kircher, Villette and others: he was clearly interested in the burning-specula themselves, as well as the pyrognostic properties of materials they revealed.[98] Soon Priestley was testing the powers of an 'incomparable lens' of 12-inch diameter and 20-inch focal length, bought from the London glass-manufacturer William Parker, whose company generously provided many other pieces of glass apparatus over the next three decades.[99] Priestley admitted that 'without this very instrument' he did not know how his experiments with air 'could have been made at all; certainly not to so much satisfaction'.[100] In August 1774, probably on the Bowood estate of his patron the Earl of Shelburne, Priestley famously used Parker's glass to heat 'calx of mercury' (mercuric oxide). The emitted vapour aided the combustion and the respiration of mice, convincing Priestley that he had isolated 'dephlogisticated air', although his understanding of this reaction changed in later works and was never absolutely clear.[101] Visiting Paris in October that year, he told Lavoisier about his experiment, and the rest of the story of oxygen's discovery is, of course, one of the most famous in science, although it is now generally considered that Carl Wilhelm Scheele (1742–86) had isolated the gas one or two years earlier.[102]

Priestley's correspondence over subsequent years is abundant with praise for Parker's skill and generosity, as his invention established itself as an essential tool in the chemist's laboratory. 'I am making the most of the fine sunshine we now enjoy', he explained to Josiah Wedgwood (1730–95), 'and have lately discovered some very remarkable new facts, which promise to throw much new light on the doctrine of air, &c. They could not be made but by means of a burning lens.'[103] To use the lens, however, Priestley had to adapt his apparatus and take special care, grinding the glass vessels very thinly and applying the heat gradually to avoid cracking upon their thermal expansion (and was successful where Lavoisier had previously failed).[104] Priestley subjected many different materials to the lens's heat, and his discoveries included the reduction of metallic oxides by hydrogen, and the production of carbon monoxide when burning charcoal. However, Priestley interpreted all of the experiments in accordance with the dominant theory of phlogiston formulated by Georg Ernst Stahl (1659–1734), and at first did not recognize carbon monoxide as a distinct gas, but confounded it with hydrogen (identified as 'inflammable air').[105] Yet Priestley's discovery of oxygen inspired Lavoisier to explore such reactions further, and he concluded that combustion and calcination involved not a release of some of the 'phlogiston' each substance apparently contained but a reaction between the solid substance and the respirable component of the 'common air' Priestley had found.[106]

In the Birmingham riots of July 1791, Priestley's home and laboratory were destroyed, and Parker's lens with them. After Priestley's move to Pennsylvania, Samuel Parker, son of William, supplied glassware across the Atlantic, including a replacement burning-lens.[107] The interaction between Priestley and the Parkers in testing the abilities and design of burning-glasses offers further confirmation that eighteenth-century chemistry established itself as a 'technoscience' in which connections between science and technology were, in the words of Ursula Klein, 'entrenched in a shared material culture', exchanging knowledge, skill and instruments.[108] Yet such a relationship had been shaped by social, political and cultural conditions surrounding burning-glasses for at least two centuries.

William Parker enjoyed a share of Priestley's glory, with one of his compound-lenses celebrated in Abraham Rees's *Cyclopaedia* as the 'most powerful' ever constructed, reaching a focal temperature estimated (using Wedgwood's pyrometer) to be 1096 °F (Figure 8).[109] Parker invited some of London's most learned, including Sir Joseph Banks (1743–1820), to witness his 'burning Lens of greater Diameter than any one made in England' at his house at 69 Fleet Street, on 10 July 1782. Banks could not attend, so with another invitation Parker sent experimental results, which recorded the melting of '10 Grains hammerd iron' in 17 minutes, among other feats.[110] But as this was an expensive business, and unable to find a suitable buyer at home, Parker was apparently 'induced to dispose of it' to Captain William Mackintosh for the (still relatively princely) sum of £700, and it joined the inventory of scientific instruments sent as part of Viscount George Macartney's delegation to China.[111]

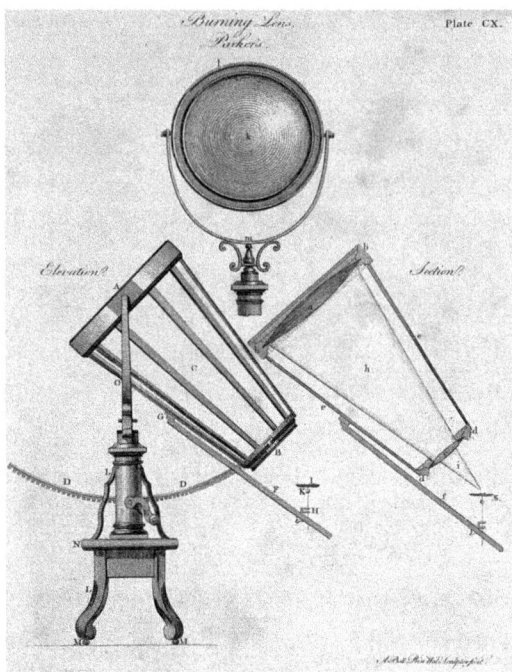

Figure 8 Andrew Bell, engraving of Parker's burning-glass, in *Encyclopaedia Britannica; or, A Dictionary of Arts and Sciences*, ed. James Millar, 4th edn, 20 vols (Edinburgh: Andrew Bell, 1810), after III, 788 (Plate CX). Wellcome Collection CC BY.

Arriving in the summer of 1793, the embassy was both a reconnaissance and a trade mission. The items transported, therefore, reflected a desire to express British commercial superiority as a product of Enlightenment values and knowledge.[112] Finding the lens to be 'extraordinary' and 'an object of singular curiosity', Macartney (1737–1806) records in his journal a concern that it must be demonstrated appropriately, since its aesthetic value does not do justice to its instrumental one: 'If it fell into the hands of the Chinese merchants and were presented through their channel to the Emperor it might tend towards the disparagement of our fine things.' At least at first, Macartney's apprehension proved unfounded, in that 'Parker's great lens … seemed to strike the Chinese in a most particular manner' when first shown to them, not in operation. However, the witnesses 'shook their heads' when they learnt the lens could not be reproduced easily, and also 'expressed the utmost astonishment' when told it would take time to set up.[113]

The embassy included the Scottish schoolmaster and lecturer James Dinwiddie (1748–1816), who recorded the demonstration of Parker's glass:

Wood set on fire, even the Chinese cash melted by the power of this apparatus, seemed to excite no other feelings in the prime minister than lighting his pipe at the focus in derision of its usefulness. A little mirth was occasioned by a eunuch who had more curiosity to approach the focus with his finder, and felt more than he had desired for.

It was taken away by Emperor Hongli's men the next day, and Dinwiddie bewailed that the lens 'of which there is not an equal in the world – is consigned to everlasting oblivion'.[114] Rees's *Cyclopaedia* also reports that it was left 'in the hands of persons, who most probably know neither its value nor use'. Yet the sense of discomfort and frustration at the Chinese response was perhaps because its flippancy exposed (in its own way) the flaws of a fragile, intermittently working, and poorly purposed technology, and ultimately served to deflate the hubris of the British Enlightenment.

An iconography of Priestley with the burning-glass developed during the nineteenth century, and produced a more successful display of British science. Many of the caricatures of Priestley in his own lifetime, the most famous probably being 'Dr. Phlogiston' (July 1791), had associated his inflammatory radicalism with hot 'French spirits' and/or gunpowder, adapting Priestley's metaphor that dissenters were 'laying gunpowder, grain by grain, under the old building of error and superstition'.[115] Long after animosity towards Priestley had boiled over in the form of the Birmingham riots, the Victorians came to appreciate him as a scientific pioneer, and as part of this cultural rehabilitation his visual depiction transformed. On the centenary of Priestley's isolation of oxygen, Birmingham performed a 'great act of retributive justice' to his 'illustrious memory' by erecting a marble statue which formed a stark contrast to the explosive caricatures that blighted Priestley during his lifetime (Figure 9). The *Birmingham Post* for 8 August 1874 explained that Priestley is represented

> holding in one hand a 'burning lens', and in the other a tube, dipped in a vessel containing mercury, ... The incident represented is the philosopher's great discovery of oxygen, and the importance of the discovery is indicated by the attentive, or even the intense, gaze with which he regards the action of the sun's rays.[116]

The sculptor Francis John Williamson (1833–1920) had reduced the burning-glass down, equipping Priestley, Prometheus-like, with the power of the Sun

Figure 9 F., wood engraving of statue of Priestley at Birmingham, after Francis John Williamson, 1874. Wellcome Collection CC BY.

in his hand. In capturing this foundational experimental moment, the statue evoked an image of instrumentality associated with Enlightenment rationality, heroically placing the scientist with the technological means to knowledge, and connecting Priestley's philosophical focus with the intensity of the Sun's beams. Indeed, in his speech at the statue's unveiling, the local preacher George Dawson (1821–76) reinterpreted Priestley's hermeneutics with reference to this image: 'He looked upon writings as so much phenomena to be examined by his reason, that the logic of St. Peter, as well as the peculiar preparations of Mercury, were alike to be exposed to the operations of his burning-glass, and he had to wait for the results'.[117] While the sculpture had diminished the spatial scale, by reducing the burning-glass down to aesthetic proportions, at least one viewer appreciated the temporal realism it attempted to convey, of heat (and thereby knowledge) generated methodically by the concentrated light of reason, and not in an instantaneous flash.

This sculptural tribute can be placed within what Christine MacLeod has called the 'statumania' of the mid-nineteenth century, when Britain sought to enshrine in marble its recent scientific and technological successes.

Birmingham's Priestley statue was part of a national pantheon of 'great men' glorified through civic art, many of them depicted with a career-defining object, including Josiah Wedgwood by Edward Davis (1862) in Stoke-on-Trent, holding the Portland vase; James Watt by Alexander Munro (1868) in Birmingham, resting one hand on a steam-engine cylinder; and Humphry Davy, by W. J. and T. Wills (1872) in Penzance, with his right hand on the safety lamp.[118] Each of these men had invented the object they were immortalized with, while Priestley was merely the ingenious user of Parker's glass. Priestley's reputation was also more in need of redemption. The representational demands were, therefore, different, with Priestley depicted in the intense activity of experimental observation, the others standing triumphantly with the products of their invention, and gazing outwardly to the spectator. The narrative aspect of this sculpture has been noted as very unusual for its time,[119] and one might argue that the focus upon this single episode, rather than the man more generally, perhaps suggests continued discomfort with the more radical facets of Priestley's life and career. Portraying the experiment as a kind of orchestrated sublimity, the sculpture rewrites this historical moment of discovery, and the 'aesthetic proportions' which reduce the instrument at Priestley's command have the effect of elevating and focusing upon the scale of his particular personal achievement. The size of the lens makes the experiment seem so simple, and yet almost paradoxically renders Priestley's accomplishment even more impressive, that with such a rudimentary tool he can make such an important finding: obscuring Parker's intricate and much larger device, the artisanal and intellectual labour involved in its construction, and burning away the contested history of oxygen's discovery, setting in stone Priestley's priority over the efforts of Scheele and Lavoisier. In subsequent years, similar statues of Priestley with a handheld glass, by Alfred Drury (1856–1944) and Frances Darlington (1840–1940), were unveiled in Leeds (1903) and Priestley's hometown of Birstall (1912), ensuring that this instrument became an integral part of the refashioning of Priestley's identity.

The foundation of this image, however, was laid by Priestley himself, in Enlightenment and Unitarian rhetoric which self-consciously associated the production and manipulation of light and heat with metaphors of scientific, social and political progress led by human reason.[120] Priestley saw his diverse concerns as part of the same, interconnected programme, believing that the recent 'amazing improvements in natural knowledge' would be 'instrumental in bringing about other changes in the state of the world'. This 'rapid progress of knowledge', which is 'like the progress of a *wave* of the sea, or *sound*, or of *light* from the sun', is the means of 'extirpating *all* error and prejudice'.[121] Elsewhere, Priestley declares the 'endless progress' of the 'bounds of light' to be 'a prospect truly sublime and glorious', co-opting for the radicals what had once been the symbol of monarchical power.[122] Anna Laetitia Barbauld (1743–1825), Priestley's friend and fellow dissenter, would later poeticize this heroic image of the 'animating ray' of reason and liberty removing superstition and tyranny across regions of the Earth in her prophetic *Eighteen Hundred and Eleven* (1812). The transformative social and economic effects of this progress are evident in 'Celtic nations' as if 'some playful child the mirror turns, / Now here now there the moving lustre burns'.[123] As abolitionist discourse pointed out, however, the 'bright intellectual Sun' of liberty had yet to 'irradiate *all* the earth', with Hannah More (1745–1833) asking: 'While Britain basks in thy full blaze of light, / Why lies sad Afric quench'd in total night?'[124] These tropes of illumination would pervade the early nineteenth century and have a bearing upon the cultural experience of the Sun's power, as the next chapter will explore.

Heavenly machines

At the end of the eighteenth century, emulating Archimedes's thermal optics was suggested as not only a scientific but also a military endeavour, returning to the myth full circle. The Belgian Étienne-Gaspard Robert (1763–1837) would become famous (as 'Robertson') for his magic lantern shows and ballooning exploits in the late 1790s.[125] During the Terror he returned to his hometown of Liège and there made the acquaintance of François Villette's son. Robertson's *Memoirs* describe how their conversations instilled his 'passion' for optical science, and he was astonished that the effects of Villette's mirror had not inspired military or industrial application.[126] Robertson identified that one particular reason for this neglect was the mirror's restricted motion, and he took it upon himself to construct a new device whose position could be modified at ease, unlike both Villette's and Buffon's (p. 128). Robertson's design was very different to Villette's single concave, instead resembling Buffon's array of flat mirrors, but with a thousand of them (p. 126). Yet Robertson emphasized the grand simplicity of his design over Buffon's: dispensing with springs, he adopted Archimedes's screw as the mechanism to control his mirrors (pp. 133–4). According to Robertson, in all probability the Syracusan genius had himself used this mechanism within his fabled mirror device (p. 133), and this combination of Archimedean technologies, therefore, acted as further proof of the story's authenticity. Declaring that Archimedes's mirror had been reinvented (p. 134), Robertson – like many others before him – engaged in a mutual validation of the myth and his own techno-scientific credibility. And like Archimedes, Robertson sought to defend his country from maritime invasion, although rather than annihilating the ships completely he had the more modest ambition of burning their ropes (pp. 135-36).

In 1796, Robertson was ready to make his proposal public, but needed military protection to enable his safe passage to Paris, and so divulged his project to war leaders including the brigadier general Gilles Jean Marie Roland de Barazer, Chevalier de Kermorvan (1740–1817), who was inspired to dream of all enemies of the Republic being 'réduits en cendre' (reduced to ash) (p. 134). Once in Paris, the directoire appointed three commissioners to examine Robertson's plans, including Louis-Bernard Guyton, Baron de Morveau (1737–1816) (p. 138), who had successfully used a burning-lens to combust diamond.[127] Robertson attended a session of the Académie des Sciences with a model of the machine, producing astonishment among the members (p. 142). This was not enough to secure the considerable investment required in taking the idea forward, and the military scheme never came to fruition. However, as Robertson directed his efforts into other areas, his previous thermal optical research would not be in vain. He had learnt in detail, from written accounts and the oral testimony of Villette's son, of the images of monstrosity and horror produced by Villette's burning-mirrors (pp. 114-16). Realizing their theatrical potential, his own mechanized mirrors formed a crucial part of his phantasmagorical entertainments, zooming in on his magic lantern projections to create chilling images that horrified and delighted the audiences of Paris in equal measure.[128] As Robertson's *Fantasmagorie* seemingly conjured up the apparitions of those fallen during the bloody days of the Revolution, the spirit of Villette's technologies lived on.

While Robertson's vision of a French military assemblage of mirrors ready to burn an invading British fleet had been cast aside by the Directoire, somehow the idea made its way across the channel, and was duly taken up within the fantasies of British caricaturists, for whom the burning-glass became a brilliant instrument of political satire. After the Treaty of Amiens collapsed, during 1803 invasion by French forces was an almost daily fear, and the graphic satires of this time developed a patriotic iconography of defence that usually featured encounters with Napoleon Bonaparte (1769–1821) in the channel. James Gillray (1756–1815) reduced the First Consul to 'Little Boney', but this satiric miniature became a standard motif, often in combination with a Brobdingnagian George III (1738–1820), who needs a spy glass to see his enemy.[129] The anonymous, mock-heroic and Gillray-esque cartoon 'An Experiment with a Burning Glass' (15 September 1803) united this emblematic formula with a burning-lens conceit, depicting the diminutive Bonaparte lying in a specimen dish, with the king as a natural philosopher, burning him using a handheld glass and saying 'I think, my little fellow, you have now experienced a sing[e]ing *à l'Anglaise*' (Figure 10).[130] This piece is strikingly similar to 'A British Chymist Analizing a Corsican Earth Worm!!', published by William Holland in July 1803, in which George inspects,

An Experiment with a Burning Glass

Published by F Pask & Vigor Lane near Sackville St Sep. 15 1803

Figure 10 'An Experiment with a Burning Glass' (London, 15 September 1803). Library of Congress.

with a magnifying glass, the worm of Bonaparte in a chemist's retort.[131] 'An Experiment with a Burning Glass', however, takes the 'analysis' one crucial stage further: moving from observation to combustion, and turning scale-driven ridicule into murder-fantasy.

Soon afterwards, the hand-coloured etching 'An Attempt to destroy the British Fleet' (1803), published by Piercy Roberts, seemed to refer more directly to Robertson's idea, placing a glass in the hand of Napoleon who, while focusing the rays over the channel at the approaching ships, laments 'Fire and gunpowder! They told me a good burning-glass would do the business in an instant, but it has no more effect than if I threw a snowball at them' (Figure 11). Although it is unclear which satiric assault came first, this particular caricature seems to be in dialogue with the French 'Vent contraire' ('Headwind'), published by Aaron Martinet in November–December 1803, which shows a group of gentlewomen on the British coast fruitlessly waving their fans to prevent the landing of the French fleet.[132] These mirror images of one another seem to trade on the sense of frustration on each side that significant technologized

Figure 11 'An Attempt to Destroy the British Fleet' (London, 1803/4). BM 1985,0119.414. © The Trustees of the British Museum.

answers to defence were not forthcoming: the most significant large-scale innovation of the Wars was the mass production (at Portsmouth) of wooden pulley blocks, needed on sailing ships;[133] but hardly the ultimate weapon some might have hoped for.

Both 'An Experiment with a Burning Glass' and 'An Attempt to destroy the British Fleet' capitalize mock-heroically on the dislocation of scale between the handheld and the large glasses, foregrounding the semiotic relations between the two optical objects. Disproportion is, of course, a key rhetorical weapon of ridicule, and the jokes about Napoleon's stature, combined with the fluid relational meaning between the two scales of glasses (and their ocular and thermal usages), produce irresistible caricatures that elevate the magnifying glass to military solution. The two caricatures also patriotically align the French with ignorance of pyrogenic optics, and the English with its mastery, reversing what had long been the real state of affairs (Parker excepted).

The satiric value of Napoleonic burning-glass weapons was further exploited in 1811, when a contributor to the radical *Belfast Monthly Magazine* proposed ironically that the British should unite the whole nation's domestic mirrors against France, with the only material inconvenience being that 'men would be obliged to go unshaved'.[134] The putative author 'Brutus', a name associated with tyrannicide (and, therefore, a common revolutionary symbol of the period),[135] first contemplates how thrilled Thomas Malthus (1766–1834) and his 'disciples' would be at the 'occasional use of this great machine' in controlling population growth. Here the author is possibly recalling Malthus's simile that 'the French revolution, … like a blazing comet, seems destined either to inspire with fresh life and vigour, or to scorch up and destroy', seeking to imply that the philosopher enjoyed the thought of thermal genocide.[136] The letter's ironies continue as the patriotic 'Brutus' himself revels in the 'glorious sight' of the destruction that would be caused by directing the rays first at Paris ('that hotbed of vice'), leaving some time for the French to be 'frightened, and see the folly of all their grand schemes of conquest, before … exert[ing] the full powers of our heavenly machine' over the whole country (p. 374). 'Brutus' is, therefore, a kind of Modest Proposer, adopting the faux civility of Swift's similarly violent narrator, and taking the paradoxical symbolism of the burning-mirror as rational destroyer of things to its *reductio ad absurdum*.

The political satire becomes more specific as 'Brutus' recognizes the potential pageantry of this scheme, imagining the Prime Minister Spencer

Perceval (1762–1812) personally aiming the death ray, with all the bishops assembled and 'a stage ... erected for the Duke of York; [who] will delight to see Dunkirk in flames': poking fun at Prince Augustus Frederick, Duke of York and Albany (1763–1827), whose unsuccessful siege of that town in 1793 had led to considerable public humiliation.[137] Not only satirizing Brutus's unsavoury politics and morality, the parody also ridicules his ignorance of the practical difficulties involved, since he believes 'there is little doubt but a few hours of bright sun-shine will for ever prevent any trouble from France'. Yet for all of the letter's radical, anti-war satire, its ending suggests the author's attraction to the imaginative potency of annihilation, as it fantasizes how France's 'forests destined for future navies, and her most splendid palaces will soon be transformed into gas, and the whole foundations of the empire with her 25 millions of inhabitants be reduced to perfect scoria' (p. 374). 'Gas' is a fitting final state for the enemy across the water: a term that, particularly following Edmund Burke's lament that 'wild *gas* ... is plainly broke loose', had become associated pejoratively with France's political volatility, after the rationalization of chemical nomenclature.[138]

Over the eighteenth century, the technology of the burning-glass had become a supreme mediator of not only the Sun's power but also expertise between philosophers and inventors, and knowledge between the developing scientific institutions and popular audiences of natural philosophy. At the century's close, large burning-lenses were for Baron de Morveau and Humphry Davy (1778–1829) instrumental in their discoveries that diamond and charcoal were each made of pure carbon.[139] Despite these achievements, advances in furnace construction, particularly the portable design of Joseph Black (1728–99), by that time meant that burning-glasses – expensive, fragile and intermittently functioning items – started to be used less and less as laboratory tools (in 1782, for instance, Lavoisier managed to melt platinum by burning charcoal in a blast of oxygen).[140] Nevertheless, the cultural life of burning-mirrors brings into focus the experience and symbolic power of the Enlightenment. The construction and artistic reproduction of these solar technologies crossed an array of philosophical, disciplinary, professional, social and geographical boundaries, and – as beams of truth and reason seemed to pour forth from them – these objects of knowledge appeared to reveal the coexistence of ethereal and material worlds.

Feeling the Promethean heat: Romantic radiance and the power of invisible light

Combustible minds

Benjamin Robert Haydon (1786–1846), the painter and friend of Wordsworth (and later, Keats), had for several years, and for up to 14 hours a day, studied the marble statues in Lord Elgin's backyard on Park Lane.[1] But on 10 November 1815 he experienced a new kind of creative sensation in relation to one of the friezes:

> As I looked at the Lapitha who grapples a Centaur, I dwelt on it with more intensity than ever. Its beauty, its divinity, came over my soul like the influence of an angelic spirit, … It felt as if a supernatural being was directing the beam of a burning lens to fire my soul. … I imagined the red & fiery point was burning my heart, & then, … suddenly sprung up in my feelings with an elasticity of Spirit, as if one had slipped out of one's skin. All night its Divinity has beamed to my brain.[2]

Haydon fantasizes a quasi-spiritual metamorphosis and perhaps total physical annihilation in the purifying, combustive glory of sublime (and sexually charged) art.[3] In so doing, he appropriates a cluster of connotations surrounding the burning-glass, whose figurative potency ensured its presence within the Romantics' technological imaginary. This instrument might be dramatized as an instrument of not only physical but also metaphysical and psychological transformation or discovery. It could stand metaphorically for intensity, precision, and concentration of thought or feelings, and was envisaged as an instrument of divine power, with the heart as the receiver of God's radiant love (as we saw particularly in Chapter 1). As the previous chapter explored, from the mid-eighteenth century these solar technologies of light and heat, and the knowledge they generated, were celebrated as compelling symbols of the Enlightenment and literal embodiments of the light of reason, as Priestley, Lavoisier and others used them to facilitate chemical breakthroughs. No wonder then, that Friedrich Schiller (1759–1805) figured burning-glasses as agents of discovery in his 'Ode to Joy' (1785): 'Aus der Wahrheit Feuerspiegel / Lächelt sie den Forscher an' ['From the truth fire-mirror / She [Joy] smiles at the researcher'].[4] Their abilities enshrined the paradox that the production of knowledge or diffusion of truth might involve combustion and with it a kind of destruction or, at least, change of state. Light and heat were both creators and destroyers – characteristics associated with supernatural beings both Judaeo-Christian and Classical – and these optical technologies became analogues for their divine powers.

Haydon's account reveals, however, how the Romantics also appropriated the burning-glass as an analogue of the imagination, and used it to frame their turn inward. From the action of this figurative burning-lens, Haydon feels as if he 'slipped out' of his skin, changing his state to become an immaterial thing, analogous to the evaporation of substances and generation of new knowledge brought about through the application of heat in pneumatic chemistry. The focus of feeling shifts throughout: first it is the 'soul', then the 'heart', then finally 'beamed into his brain': a material conception of the mind in which there is a kind of telepathic, proto-spiritualist transfer via invisible rays of heat (and one which obscures the erotic potential of Haydon's experience). It is notable, however, that this translation of meaning from artwork to himself is triggered by his own initial gaze upon the frieze, suggesting a constructivist idea of perception. The subject is 'creator and receiver both', to quote *The Prelude*: a motif we will encounter in other thermally figured experiences of the period.[5] In engaging with solar energy as a way of rationalizing their thoughts and feelings, the Romantics were not just relying upon old tropes, as this chapter will show. New research, particularly from William Herschel (1738–1822) and Humphry Davy (1778–1829), both stemmed from and stimulated new ways of thinking about the Sun and its power. Consequently, these influenced ideas of environmental sensibility and natural theology, at a time when humanity's place within 'nature' became an important intellectual crux. Romanticism has been characterized as a movement that instilled an acutely ecological consciousness.[6] Yet Romantic encounters with solar energy, harnessing the ultimate source of physical and imaginative creation, arguably intensified the Enlightenment's instrumentalized notions of nature, perpetuating anthropocentric attitudes.

The relationship between human, nature and the divine was a central preoccupation of Johann Wolfgang von Goethe (1749–1832), and this concern often manifested itself in optically inflected discourse. In his memoirs, Goethe recalls with some amusement that as a youth he became disillusioned with the 'dry course of morality' of his family's Protestant instruction. Instead, he sought 'immediate communication with the great God of Nature', and 'after the manner of the patriarchs' built himself an 'altar':

> A flame was to arise, figurative of the human soul ascending towards its Creator. … My father had a handsome music-desk of red-lacquer, adorned with golden flowers, in form of a four-sided pyramid, … [I] laid my specimens of natural history upon it in gradation, … [The sun] rose high enough to allow me, by means of a burning glass, to light my grains of incense, scientifically arranged on a fine porcelain cup. … I wished to repeat my pious ceremony. Unluckily, when the sun appeared I had no porcelain cup at hand; … The grains of incense, in burning, had covered the fine red lacquer, and the gold flowers, with black spots; as if the evil spirit, driven away by my prayers, had left the indelible traces of his feet on the desk. The young pontiff now found himself in sad perplexity. … he never afterwards had the courage to attempt to repeat his sacrifice, and he thought he saw in this accident, a warning of the danger of attempting to approach the Deity.[7]

This juvenile act of Enlightenment natural theology combines parodically a kind of ancient Egyptian or Babylonian ritual with chemical experiment, in an uncannily domestic setting. Assembled for destruction, the specimens are placed in 'gradation', in some form of taxonomy, ordering nature according to human reason, while the incense has been arranged on the cup apparently 'scientifically'. But the aim of this pyrognostic procedure is spiritual enhancement, not intellectual progress, facilitating analysis of the self, not of external substances, through seeking closer proximity with a Pantheistic Creator equated with the Sun. Yet, functioning as a kind of cautionary tale, the young Goethe learns the hard way that one cannot control the Sun oneself: when it arrives one must be prepared, or risk the wrath of an 'evil spirit'. Moving into the third person to describe the moment of his despondency, Goethe is a 'young pontiff', connoting in his sacred ministration not only a kind of Catholic rebellion against his family's Protestantism but also the Latin derivation 'pontifex', or 'bridge-builder', as he attempts to connect the realms of the natural and divine.[8]

The young Goethe's act crystallizes the blurred boundaries between worship and study, revelation and discovery, showing that even the simplest technological manipulation of the Sun's power is coupled with a sense of sublime awe. This was an experience the deistical Goethe would draw upon when imagining the thoughts of the youthful Prometheus:

> I turned my wandering gaze
> Up toward the sun, as if with him
> There were an ear to hear my wailings,
> A heart, like mine,
> To feel compassion for distress.[9]

Goethe characterizes the anthropomorphic deification of the Sun as an almost innate feeling of the juvenile human, who seeks sympathetic encounter with an elusive, tangibly unknowable heavenly body which seems the closest one can come to divine presence. Yet Goethe's veneration for the Sun declined very little as he matured, perhaps since he believed that 'the highest achievement of the human being … is to have probed what is knowable and quietly to revere what is unknowable'.[10]

Goethe's obsession with optics became substantially more sophisticated over the next half-century. Most significantly, Goethe constructed an alternative theory of colour to Newton's: one emphasizing human perception over physical reality, and claiming darkness to be not just the absence of light, but a thing in itself. The close correspondence and creative affinity between the self and the solar body (and, by corollary, man and God) manifested itself again in Goethe's translation of Plotinus (AD c.204–270): 'Were the eye not of the sun, / How could we behold the light?'[11] Claiming to have 'discovered light in its purity and truth', Goethe considered it his 'duty to fight for it' – an ambition he maintained in his drama and fiction.[12] Goethe's youthful sacrifice to the Sun anticipates Faust's praise of dawn, while the transformative power of light features in *The Lily and the Green Snake*, when Midas-like the beam of the ferryman's lamp turns all it touches into gold.[13] The burning-glass, meanwhile, had a unique metaphorical appeal in Goethe's eyes, because of its ability to unite and concentrate: 'Form … is, once for all, the burning glass, through which we draw the scattered rays of holy nature to the hearts of men in a fiery focus'.[14] The convergence of the rays is, of course, not discernible itself, only observable in its effects, making this phenomenon particularly appealing for the generation of metaphor. But to the deist Goethe, the burning-glass had more than a tropic value: it manifested the possibility that the Sun was a reflection of God's power and benevolence. Even Goethe's last demand, on his deathbed, was reputedly to 'Open the shutters, that more light may enter'.[15]

The discourse of light had a different presence within the work of William Cowper (1731–1800), concurrent with his evangelical Christianity.[16] Yet his similar adoption of the burning-glass as a conceptual tool hints at the important place of technology within Romantic conceptions of mind. Cowper's self-conscious 'Conversation' (1782) poem advises:

> A tale should be judicious, clear, succinct,
> …
> And new or old, still hasten to a close,
> There centring in a focus, round and neat,
> Let all your rays of information meet: (ll. 235, 238-40; *Poems*, I, 360)

One's ideal narrative should work like a burning-glass, carefully converging many beams for its skilful climax: an aesthetic aspiration Cowper puts into practice in one of his early love poems, written for his cousin Miss Theodora Jane Cowper. Entitled simply 'A Song', the poem not only absorbs many of the established burning-glass tropes but also builds this instrument into its very structure:

> The sparkling eye, the mantling cheek,
> The polish'd front, the snowy neck,
>> …
>> All meet in you, and you alone.
>
> Beauty, like other pow'rs, maintains
> Her empire, and by *union* reigns;
>> Each single feature faintly warms:
> But where at once we view display'd
> Unblemish'd grace, the perfect maid
>> Our eyes, our ears, our heart alarms.
>
> So when on earth the God of day
> Obliquely sheds his temper'd ray,
>> Through convex orb the beams transmit;

The beams that gently warm'd before,
Collected, gently warm no more,
But glow with more prevailing heat. (ll. 1-2, 6-18; *Poems*, I, 13-14)

The idea of union seems to have influenced the choice of 'Spanish sestet' or 'sextilla' as the verse form (AABCCB), with each stanza pulling back to focus upon its centre. Not only are Theodora and the personified 'Beauty' figured as burning-glasses concentrating individual features together, but also the very act of the imagination on the part of the gazing speaker in uniting these images corresponds to the operation of a 'convex orb', the poem ending with his passionate energies released as 'heat'. 'A Song' puts itself forward as a medium that focuses disparate parts of the world into an intense spark.

Cowper's poem shows us that the burning-glass should be located within the cluster of metaphors that, according to M. H. Abrams, reveals an aesthetic shift from Platonic conceptions of the mind's passivity, to the characteristically 'Romantic' view (ultimately derived from Plotinus) of the active imagination. Yet the intellectual history of the burning-glass problematizes Abrams's schema, in which popular epistemology seemed to move from mirror metaphors to those of the lamp: from imitation to expression.[17] The *burning*-mirror's parabolic reflections (or burning-lens's refractions) actively produce a focus that can itself create and destroy rather than inertly receive. As a mechanical metaphor of mind, the burning-mirror or lens suggests an imagination with its source in the external world, but which in the act of concentration produces something of greater intensity. Both Joseph Priestley (1733–1804) and Samuel Taylor Coleridge (1772–1834) invoke the burning-glass metaphor implicitly in this way, by describing (respectively) that the 'close attention to a subject which composition requires, unavoidably warms the imagination: then ideas crowd upon us, the mind hastens, as it were, into the midst of things', and that 'man's mind is the very focus of all the rays of intellect which are scattered throughout the images of nature'.[18] The burning-glass is a kind of metaphorical 'missing link': it does not produce its own reciprocal, 'auxiliar light', but has a transformative capacity in relation to the radiance it receives.[19]

The responses of Haydon, Goethe and Cowper show us how the trope of the burning-glass gained new strength during the Romantic period. This was a time characterized by not only a shift from mimetic to expressive notions of artistic creativity and reception sometimes conceptualized technologically (from the mirror to the lamp, according to Abrams), but also political and social revolution often represented through images of light and heat, *and* a radical change in understanding of the Sun's rays. Issues of liberty, imagination and humanity's relationship with the environment were being debated and developed, and instrumentalized notions of the Sun, heat and light provided those issues with a new metaphorical structure. These enhanced the commonplace associations of liberty and reason with lights which encouraged life, health and knowledge, based on Platonic ideas and the image of Christ as 'light of the world' (Mt. 5.14–16).[20] Addressing the general assembly of the Piques Section of Parisian revolutionaries in October 1792, its president the Marquis de Sade (1740–1814) proclaimed: 'The power … of your representatives is like the ray of the sun, reflected by the burning glass; … which will illuminate the earth only with the fires which you have transmitted to them.'[21] For Mary Wollstonecraft (1759–97), the 'irresistible energy of the moral and political sentiments' of Voltaire, Rousseau and others had 'kindled into a blaze the illuminating rays of truth, which … completely undermined the strong holds of priestcraft and hypocrisy'. Davy apostrophized reason as a 'radiant Sun the fountain of true light', long enslaved by the 'Craft / Of Priests & Politicians'. Meanwhile, Mary Robinson (1756/8?–1800) modified the Renaissance trope of the female's combustive gaze to endow personified Liberty with a Promethean rebelliousness of 'sparkling eyes / [which] Snatch'd radiance from the sun!'.[22] Cultural, political and scientific representations, therefore, drew upon ideas of solar energy and, especially, of utilizing that power both literally and metaphorically.

Moreover, the Romantic period, as we know it, is book-ended by the emergence of new knowledge and ways of thinking about solar heat – the discovery of 'invisible light' (i.e. infrared radiation) beyond the visible spectrum and the measurement of the immense solar influx upon the Earth. Coinciding with these developments in the science of solar heat was a new profound interest in classical characters and narratives associated with the Sun, light, heat and flame, including Prometheus. There were, of course, political and aesthetic reasons for the Romantic preoccupation with these figures,[23] but their physical associations enhanced the symbolism, instilling solar power and its human wielders with revolutionary potential and heroic individualism, and framing combustion as a restorative form of change. In his defiance of Zeus, Prometheus was seen as the archetypal rebel and traitor, and yet also the hero and freedom fighter for humanity who endures agonizing, perpetual punishment for giving power to the weak. Moreover, in the combination of liberty with confinement, primitive power with technological mastery, nature with industry, radical artistic aspiration with theft/plagiarism, creation with destruction, rightful inheritance with fraudulent usurpation of forbidden knowledge,[24] Prometheus's potential for heteroglossia was shared with the solar technologies he was imagined wielding.

Several Romantics reimagined the Prometheus myth optically, replacing his fennel-stalk with a mirror or lens carrying down the fire of heaven. However, this technological reworking was far from new, and is traceable at least as far back as the late sixteenth century, when it was invoked in the culture of natural magic.[25] As we will see, in the Romantic period, the Promethean burning-glass acquired new connotations of political, social and intellectual intensity that bound up human improvement with images of scientific progress. Paralleling the fires of revolution, the combustive action of the burning-glass in experimental practice (which isolates elements from their compounds and generates new chemical knowledge) is represented as a liberating and revelatory process. Yet at the same time, this misappropriated Promethean heat is 'stolen' from nature, putting nature's own destructive force into the service of man.

The politics of the Promethean burning-glass are evident in 'Monody on the Death of the Right Hon. R. B. Sheridan, spoken at Drury Lane Theatre, London' by Lord Byron (1788–1824). Richard Brinsley Sheridan, who died on 7 July 1816 at the age of sixty-four, had been a Whig MP, playwright and sometime owner of Drury Lane Theatre. This epicentre of the London stage had been devastated by fire several times in its history, with its third incarnation burnt down after only fifteen years on 24 February 1809. Byron's poem associates Sheridan's radicalism with Drury Lane's fate, and perhaps seeks to recall *Pizarro: A Tragedy* (1799), Sheridan's translation of August von Kotzebue's *Spanier in Peru* (1796), some of which takes place in the Peruvian 'Temple of the Sun'.[26] Byron laments the loss of Sheridan's dramatic and political oratory:

A mighty Spirit is eclipsed – a Power
Hath passed from day to darkness – to whose hour
Of light no likeness is bequeathed – no name,
Focus at once of all the rays of Fame!
The flash of wit – the bright Intelligence,

The beam of Song – the blaze of Eloquence,
Set with their Sun, but still have left behind
The enduring produce of immortal Mind;
...

Here in their first abode you still may meet
Bright with the hues of his Promethean heat,
A Halo of the light of other days,
Which still the splendour of its orb betrays. (ll. 23-30, 55-58)[27]

Byron wrote the 'Monody' around the same time as 'Prometheus' (and in the company of the Shelleys), yet the eponymous poem focuses upon punishment and suffering entirely, and pays no attention to the cause of the crime. Instead, the heroic couplet 'Monody' is the place in which Byron contemplates the possibilities of 'Promethean heat'. Sheridan is an unrivalled light, whose fame was a result of his collected talents and who, Prometheus-like, used his creative fire to bring characters vividly to life on stage. Sheridan's heat also brings to mind his radical politics, which targeted such colonial tyrants as Warren Hastings.[28] The monody, an elegiac song performed by a single voice, is an appropriate, concentrated form through which to celebrate the many talents of a single man whose 'Power / Hath passed from day to darkness', with Byron here echoing Wordsworth's elegy on the death of another Whig, Charles James Fox (1749–1806).[29] Significantly, however, this light and heat is but a memory, and Byron wrote the poem during the 'year without a summer', when his experiences of the hot Sun itself seemed to be relegated to the past (to which we will return).

Dissecting the sunbeams

While Byron was drawing upon technological tropes of a certain vintage, knowledge of the Sun's rays themselves had moved rapidly forward by this time. In a series of studies conducted in the 1790s and published in the *Philosophical Transactions of the Royal Society*, William Herschel (1738–1822) pioneered what we now call astronomical spectroscopy. Since Newton, sunlight had been thought of as 'a mixture of several sorts of coloured rays',[30] but Herschel made the astonishing discovery of radiation which existed beyond the visible, forever changing our relationship with the Sun. Herschel's first paper on the topic began by challenging the assumption made by natural philosophers and poets alike regarding the concentration of solar rays:

> In the focus of a burning lens, it seems to be natural to suppose, that every one of the united rays contributes its proportional share to the intensity of the heat which is produced; and we should probably think it highly absurd, if it were asserted that many of them had but little concern in the combustion, or vitrification, which follows, when an object is put into that focus.[31]

Herschel, using 'various combinations of differently coloured darkening glasses', discovered that 'the power of heating and illuminating objects … might not be equally distributed among the variously coloured rays' (pp. 256, 255). With some glasses he 'felt a sensation of heat, though I had but little light; while others gave me much light, with scarce any sensation of heat' (p. 256). Employing a prism to refract light of different colours, which were then isolated through horizontal slits in a pasteboard screen and aimed at a set of thermometers, he was able to map the variation in heat and light across the visible spectrum, concluding that red rays were the hottest (p. 262).[32] Crucially, Herschel showed that radiant heat consisted, at least partly, of 'invisible light' beyond the red: 'Rays … that have such a momentum as to be unfit for vision'.[33] He even speculated that the rays in this 'invisible thermometrical spectrum' might exceed the visible ones in number.[34] Technological manipulation of the invisible rays then became a crucial area of research, as Herschel employed a concave mirror and convex burning-lens to show that the invisible heat obeyed laws of reflection and refraction similar to those of light (see Figure 12).[35]

Having discovered Uranus in 1781, with his 'Dark' or 'invisible heat' Herschel managed to astound the scientific community again. The Royal Society's president Sir Joseph Banks (1743–1820) declared: '[As] highly as I prized the discovery of a new planet I consider the separation of heat from light as a discovery pregnant with more important additions to science.' He also reported that Benjamin Thompson, Count Rumford (1753–1814), who in 1798 had shown that heat was not a substance, thought this discovery 'the most important since S^r I Newtons death'.[36] Indeed, Herschel succeeded Newton as poetry's archetypal natural philosopher, who Anna Laetitia Barbauld (1743–1825) no doubt has in mind when imagining the 'reasoning sage' who possesses the ability to 'dissect a sunbeam, count the stars, / And measure distant worlds'.[37] For Banks, the distinction between heat and light

Figure 12 William Herschel, 'Experiments on the Solar, and on the Terrestrial Rays that Occasion Heat; with a Comparative View of the Laws to Which Light and Heat, or Rather the Rays Which Occasion Them, are Subject, in Order to Determine Whether They are the Same, or Different. Part I', *Philosophical Transactions*, 90 (1800), 293–326 (Plate XIV).

'seems to solve the paradox as it hitherto has been of the moons rays being quite void of heat': the Moon's 'fire that is not brightness' as described in Percy Shelley's *Prometheus Unbound*.[38] Its application would go beyond astronomy, since 'this [separation] has been suspected by those who have considered the light of Glow worm'.[39] Herschel himself recognized there might be wider implications of his discovery; specifically that his experiments may 'clear up our conceptions about caloric', that heat was a kind of fluid or gas which moved between bodies.[40] Indeed, Banks was keen to suggest that Herschel 'use[d] the term Radiant heat instead of Caloric', since 'the French System of Chemistry ... is Likely very Soon to be subverted' due to the work of Rumford and others.[41] Banks could see the potential for Herschel's work to be appropriated again within a nationalistic assertion of British scientific endeavour, and in so doing he transformed the language of heat.

Herschel's work on 'invisible rays' encountered some early opposition,[42] but nevertheless it spearheaded research which over the course of the next century found new forms of radiation across what would become known as the electromagnetic spectrum. Indeed, its inspirational methodology did not take long to bear fruit. In 1801, Johann Wilhelm Ritter (1776–1810) and William Hyde Wollaston (1766–1828) each discovered 'invisible rays beyond the violet' end of the spectrum which had an oxidating '*chemical* agency' able to blacken silver chloride almost instantaneously (radiation which Herschel had not detected using a thermometer).[43] Ritter, believing in the fundamental dualism of nature as suggested by Friedrich Schelling's (1775–1854) *Naturphilosophie*, was convinced that light must have a kind of polarity, and deduced from the presence of invisible rays beyond the red that similar rays exist beyond the violet. Knowing that silver chloride was affected at the violet end of the visible spectrum, Ritter used this compound as a kind of detector.[44] Herschel's work was also appropriated by Thomas Young in order to support wave theory, finding it 'highly probable that light differs from heat only in the frequency of its undulations'.[45]

This was an era deeply invested in the imaginative nourishment generated by optical wonder, of natural phenomena such as the Brocken spectre, aurora and parhelion, and of artificial marvels such as the Claude glass, David Brewster's kaleidoscope and Étienne-Gaspard Robertson's phantasmagoria.[46] While some of these spectral wonders were associated with diabolical occurrences,[47] Herschel showed that light contained its own uncanny, invisible other; a thing always present, but only now discovered; a shadow self which appealed to a gothic sensibility, allowing Shelley to imagine 'rays of gloom / Dart[ing] round' Demogorgon in *Prometheus Unbound*. At the other end of the century, H. G. Wells (1866–1946) drew upon this supernatural quality in *The War of the Worlds* (1897), with the Martians' Heat-Ray described as the 'ghost of a beam of light' (see Chapter 4).[48] Probing a new invisible realm further confirmed the power of Enlightenment science, showing the supreme ability of the artificial instrument in transcending the human senses, with temperature as a new medium of vision. Natural philosophy could now 'mock the invisible world with its own shadows', as *Frankenstein*'s Professor Waldman boasts (channelling the charismatic natural philosopher and lecturer Davy).[49]

Herschel's discovery coincides with significant poetic investment in notions of the invisible more generally, as the Romantics sought to understand their profound experiences with seemingly intangible things in the external world.[50] Romantic instances of invisibility are not exclusively but often associated with the power of light: since both the light and the invisible are numinous, quasi-divine domains and because they evoke Christian ideas of the divine bringing illumination to the dark and the unseen.[51] These associations are perhaps most prominent in *The Prelude*, where the poetic imagination is 'fit / To hold communion with the invisible world', and 'the light of sense' acts like a kind of detector which 'Goes out in flashes that have shewn to us / The invisible world'.[52] Herschel's experimental endeavours in detecting infrared are, therefore, in some way analogous to poets' attempts to draw upon the invisible, and are framed figuratively as attempts to discover and harness different forms of light. Both practices are kinds of anagogic process, reaching towards quasi-divine or pantheistic creative powers. Gillian Beer has argued that theories of the ether in the late nineteenth century triggered an increasing secularization of the invisible, when it became 'a domain ... described predominantly in scientific terms'.[53] But the scientizing of the invisible had started substantially earlier. Of course, magnetism and universal gravitation were invisible 'forces' that had already received scientific attention, but Herschel showed that light – something humans previously thought they had grasped holistically – existed in at least one form that defied ocular detection, seeming to grant scientific sanction to some of the Romantics' metaphysical speculations.

Coleridge was one of the first to appreciate Herschel's groundbreaking articles. He was probably introduced to them by his friend Davy back in June 1800, when they were spending 'much time' together, and 'invisible rays' were a frequent topic of conversation.[54] Certainly, in May 1801, around the time when Coleridge had ambitions to establish a laboratory with Wordsworth and their friend William Calvert, he requested Davy to tell him 'whether Herschel's Thermometric *Spectrum* (in the Philos. Trans.) will lead to any Revolution in the chemical Philosophy'.[55] Davy's reply is not extant, but his Royal Institution lectures fervently assert the importance of Herschel's discovery.[56] In early 1802, Coleridge attended Davy's talks, apparently to increase his 'stock' of metaphors. Coleridge's chemistry remained firmly in the imaginative and conceptual realm, appropriated as ways of thinking through his experiences and feelings. This is noticeable particularly in his notebooks, which attest to a fascination with this new kind of light, its metaphorical possibilities and its religious and philosophical significance. In August–September 1802, Coleridge was using the distinction between heat and light to convey his wish to bring together varieties of belief (recalling John Wesley's warmed heart):

'Socinianism moonlight — Methodism a Stove! O for some Sun to unite heat & ~~warm~~ Light!'[57] At around the same time, he wrote 'Hymn before Sun-rise, in the Vale of Chamouni' (published 11 September 1802), in which the 'Morning-Star / … pause[s]' over Mont Blanc.[58] As the ice melts to produce 'Unceasing thunder and eternal foam' (l. 46), the speaker's admiration for sublime natural processes turns to a reverential call for the Creator. This rapturous moment of solar heat is precipitated by the detection of a different sphere of being, as he 'worshipped the Invisible alone' (l. 16). In his 'Lectures on Literature' over a decade later, Coleridge uses 'invisible Light' as an example of the need to understand differences in degree as well as in kind, although he is sceptical about the value of this oxymoronic term to a 'stranger to chemistry'.[59] Meanwhile, in *Aids to Reflection*, Coleridge's conception of the 'invisible energy' or 'translucence' of life within plants was structured by metaphors of light and solar power that acknowledged the process of photosynthesis.[60]

Most vividly, however, for Coleridge, heat lent itself to conceptions of physical, spiritual or moral anguish, and this expression was often inflected technologically. In his notebooks, Coleridge left translations of several works by the Italian sonneteer Giambattista Marino (1569–1625), since the re-imagining of this passionate poetry of unrequited love seemed an appropriate way for Coleridge to capture his feelings for Wordsworth's sister-in-law, Sara Hutchinson (1775–1835). Thermal optics had previously provided Coleridge with the perfect metaphor for the torment of (what appeared to be) a tooth abscess, with his right cheek 'placed with admirable exactness under the focus of some invisible Burning-Glass, which concentrated all the Rays of a Tartarean Sun'.[61] In his version of Marino's 'Alla Sua Amica' (September 1808), Coleridge adds the motif of the burning glass-heart, but subverts its traditional, heavenly associations:

> Lady, to Death we're doom'd, our crime the same!
> Thou, that in me thou kindledst such fierce Heat;
> I, that my Heart did of a Sun so sweet
> The Rays concenter to so hot a flame.
> …
> Hear then our doom, in Hell as just as stern,
> Our sentence equal as our crimes ~~the same~~ conspire
> ~~Those who lived here in fires unearthly enflam'd in beauty's fire~~
> Who living ~~scorch'd himself~~ basked at Beauty's earthly Fire

In living flames eternal there must burn/—
Hell for us both fit places too supplies—
In *my* Heart thou wilt burn, I roast before thine Eyes—[62]

It was apparently next to a fire that Coleridge illicitly fell in love with Hutchinson, and fiery imagery engulfed many of his writings about her as he grappled with his intense feelings.[63] By the time Coleridge translated the sonnet, his torment was particularly fevered because he had become jealous of the friendship between Sara and her brother-in-law, and believed he saw them in bed together in December 1806.[64] Coleridge's 'Alla Sua Amica' is not only a translation but also concerned thematically with the translation of feeling and energy. As the rays focus on the speaker's heart, and the poem's intensity increases, he gradually arrives at the realization that the guilty fires of passion have transformed into the tormenting flames of Hell (reminiscent of the fate of Dante's doomed lovers).[65] In particular, Coleridge's replacement for the deleted line emphasizes the contrast between the 'earthly' fire that attracts him and the 'flames eternal' this desire will cause. The sonnet ends at a kind of fever pitch, with a vivid image of mutual destruction, as their fervent feelings reflect upon and within each other.

Optical radiance and combustion provided Coleridge with further ways of articulating his pangs of desire, as he wrote (at around the same time as the sonnet) a prose note which he hoped to work up into another translation of Marino:

> If love be the genial Sun of human nature, unkindly has he divided his rays in acting on me and [Asra]—on her poured all his Light and Splendor, & permeated my Being with his invisible Rays of Heat alone/She shines and is *cold*, as the tropic Firefly. (*Notebooks*, III, 3379)

Like Ritter, Coleridge's interest in the organic universe of *Naturphilosophie* led him to believe that 'EVERY POWER IN NATURE AND IN SPIRIT must evolve an opposite', but his figurative polarities extended this into the social as well as physical world.[66] Herschel's spectral analysis of solar radiance (perhaps combined with Aristotelian stereotypes of gendered heat) becomes in Coleridge's hands an analogy for his relationship with Sara, as he tries (but inevitably fails) to make sense of his feelings. He endures scorching passion and, by corollary, psychological agony; she receives light but remains cool. This is probably the first time these coexistent differences are assigned to Sara, and Coleridge would hereafter continue to conceptualize their relationship via notions of heat and cold.[67] But the rays of love would never be united.

Uncertain gleams

The popular interventions of Coleridge's close friend Davy (both his Royal Institution lectures and 1812 book *Elements of Chemical Philosophy*) were crucial in translating Herschel's 'invisible rays' into literary discourse. Throughout Davy's career, light remained a phenomenon of intense scientific *and* poetic inspiration. During his time in Bristol with Thomas Beddoes (1760–1808), Davy became convinced not only that light – 'A substance of the greatest importance to organic existence' – is a material thing but also that some combustible bodies contained 'small portions of light' which could be 'liberated' during oxygenation.[68] Oxygen gas itself was composed of light and oxygen, so Davy believed, calling it 'phosoxygen' ('An Essay on Heat', pp. 34, 37). Since oxygen was needed for animal respiration, Davy speculated that 'LIGHT is attracted or secreted from the blood' to the brain and nerves, and hence 'essential ... to perceptive existence' (pp. 140, 144).[69] The centrality of light to plant life had been confirmed by Jan Ingenhousz (1730–99) in 1779 and, combined with his belief in its importance to animal existence, Davy concluded light to be the 'most important part in the œconomy of the universe', administered by the suns of heaven, which are 'immense reservoirs of light destined by the great ORGANISER to diffuse over the universe organization and animation' (pp. 126, 145).[70] Light and heat, however, were 'totally distinct': light was a substance, heat was not, which radically challenged Lavoisier's theory of caloric (pp. 67, 13). Heat was, instead, 'repulsive motion' (i.e. vibration) (p. 32): a return to the conceptions of Bacon and Newton. A year later, Davy used the 'valuable discoveries of Dr Herschel' to show that light and heat do have very different characteristics.[71]

Davy's lectures accorded solar energy a special status, since its 'agency ... in nature is almost universal; and it either primarily occasions, or materially influences, all the different changes that take place upon our globe'.[72] This natural philosophical interest in light unsurprisingly fed into his poetic sensibility, nourishing intellectually the rich aesthetic imagery of illumination.[73] There was a theological character to this preoccupation, and while Davy subsequently repudiated some of the more 'enthusiastic' opinions (as he called them) in his 'Essay on Heat', his poetry in the first decade of the nineteenth century continued to follow metaphysical terrain in which light manifested the pantheistic 'One Intelligence', analogous to Coleridge's 'One Life'.[74] For instance, partly inspired by Wordsworth's 'Lines written a few miles above Tintern Abbey' (1798), Davy composed some verses on 'The Life of the Spinosist'.[75] Sent to Coleridge in 1800, the poem was revised as 'Written after Recovery from a Dangerous Illness' (1808). In this poem of cosmic witness, sunlight's vital properties are observed to be combustive 'flames of life' which metamorphose 'limpid dew' into 'the rosy flower', while (in a materialist imitation of Acts 17: 28) 'insensate dust awakes, and moves, and lives' from its stimulus.[76] Storms and subsequently 'motions of the main' are also caused by the Sun, being 'renovated forms' of sunlight (as James Hutton (1726–97) called them).[77] The 'engines of Eternal will / The One Intelligence' are, therefore, ultimately solar in origin, and equivalent to the sublimity Wordsworth found 'dwelling [in] the light of setting suns'.[78] As the manuscript version asserts, the 'Ten thousand signs of ~~kindling~~ "burning" energy' observed in the stellar sky allow man to 'feel the social flame' in communion with the cosmos.[79] Reading the poem in 1800, Coleridge no doubt found much in this poem to inform his own 'Hymn before Sun-rise' (1802).

Davy's oscillation between the various literal and metaphorical meanings of light continues elsewhere, with invisible emanations from the Sun becoming, for instance, emblems of more metaphysical forms of illumination:

A sun of which we catch uncertain gleams
In this our mortal state, but which for ever
Shines from afar, wakening the spirit of man[.][80]

Similarly, in a fragment of dialogue probably from an unfinished romance, Davy imagines a dying man contemplating the afterlife, where 'the genial warmth of the sun of immortality which has shone through this shattered frame with feeble light, shall be more permanent in the regions of bliss. I feel within me new energies. … O, benevolent Deity!'[81]

Davy acknowledged solar energy to be not only the Earth's ultimate source of life and power but also a phenomenon that could be harnessed artificially by humanity, aiding scientific progress. Pointing especially to the work of Newton, Priestley and Herschel, Davy celebrates glass as a particular triumph of human ingenuity, since without it 'the gases could never have been discovered, or their combinations ascertained; ... the sublime researches of the moderns concerning heat and light would have been wholly lost to us' (pp. 318–19). Appropriately, the nobility of light and heat as physical phenomena is reflected in the sublimity of optical experiment as a pursuit, aggrandizing the achievements of those who penetrate the innermost workings of these seemingly elusive things. Going further, Davy points to experimentation offering more than just understanding, but also the opportunity to 'interrogate nature ... as a master, active with his own instruments'.[82] Davy came to appreciate this even more when using solar technologies to make significant discoveries, with a noticeable effect upon his writing and thought. And yet, while Davy acknowledges that science has transformed humanity's relationship with nature, his poetry often reveals humble devotion to an environment in which he feels only a minor agent, if not passive participant.

Davy's hyperbolic moments in the 'Discourse Introductory' are often cited as inspiration for Professor Waldman in *Frankenstein* (1818).[83] Mary Shelley (1797–1851) was certainly reading Davy's 'Chemistry' (probably the copy of *Elements* Percy had owned since 1812) in late October 1816, during the period of *Frankenstein*'s genesis.[84] In Mary Shelley's hands, Davy's recognition of the importance of light to organic life and celebration that science supplied a 'steady light of truth' are reconfigured to illustrate arrogant ambitions. Victor's description of his discovery of the creation of life seems to indulge in Davy's Platonic and biblical images to the point of literalization: 'a sudden light broke in upon me — a light so brilliant and wondrous, yet so simple, ... Life and death appeared to me ideal bounds, which I should first break through, and pour a torrent of light into our dark world'.[85]

In October 1813, Davy embarked on a continental tour, principally to investigate volcanic landscapes and activity. Along the way, Davy and his recently acquired assistant, the young Michael Faraday (1791–1867), used facilities at numerous scientific institutions, including the 'old laboratory' of the Accademia del' Cimento in Florence. They arrived in the city in March 1814, and Faraday's journal notes the 'very powerful instrument' of the 'great burning glass of the Grand Duke of Tuscany' (Ferdinand III, 1769–1824) they saw at the Accademia.[86] Davy realized that this was perhaps an opportunity to settle for once and for all the chemical identity of diamond, whose decomposition required temperatures way beyond those achievable with a standard furnace or burning-lens.[87] As noted in the previous chapter, in the 1770s Lavoisier had used two Tschirnhaus lenses in series to burn diamond. The chalk-like precipitation from that combustion led Lavoisier to believe that 'fixed air' (carbon dioxide) had been produced. Building on this work in 1796, the English chemist Smithson Tennant (1761–1815) demonstrated that the same (or at least, similar) amount of fixed air was released when burning the same weight of diamond or of charcoal, and so he suggested that structure was perhaps the only difference between the two substances.[88] In *Elements of Chemical Philosophy* (1812), Davy held the view that there must be a small difference in composition between diamond and charcoal to account for their radically different characteristics, particularly charcoal's ability to conduct electricity.[89] Davy put this view to the test meticulously in Florence, as Faraday's journal describes:

> A glass globe … was exhausted of air and filled with very pure oxygen gas …. The diamond was supported in the centre of this globe by a rod of platinum …. The Duke's burning glass … consists of two double convex lenses distant from each other about 3½ feet. The large lens is about 14 or 15 inches in diameter, the small one about 3 inches in diameter. The instrument is fixed in the centre of a round table and is so arranged to admit of elevation or depression or any adjustment required at pleasure by means of the second lens.[90]

They spent at least eight days experimenting with the burning-lens, although on some occasions 'the sun sank too low'.[91] Despite its shortcomings, for Faraday, this 'noble instrument' was worthy of Davy's 'attentive mind [which] observed & demonstrated new facts'. Moreover, the aesthetic effects of burning the diamond were, as Faraday observed, 'striking': 'the diamond gave off intense heat & a beautifull vivid scarlet light it diminished rapidly in size and became at last a mere atom … the globe was found to contain nothing but a mixture of Carbonic and Oxygene gases', just like charcoal.[92] Davy's work, sent to Joseph Banks and published in the *Philosophical Transactions*, showed that a single element could account for many different forms (now known as allotropes), and structure ascended in importance in explaining differences in material properties.[93] Davy's experience in Florence consolidated his previous admiration for the powers of the Sun, and proved to him that solar combustion is a significant generator of knowledge, a veritable supplier of intellectual light.[94]

The solar combustion of diamond by Lavoisier, Davy and others to discover its chemical 'secrets' made its way into imaginative works, continuing the association between solar heat and the generation of knowledge. *The Giants' Causeway, A Poem* (1811), by the Presbyterian minister William Hamilton Drummond (1778–1865), celebrates how 'Heralds of nature' can

> to a point condense the scattered rays,
> Whose force more potent than the furnace blaze,
> As fire the wax, each stubborn ore commands,
> And bursts the diamond's adamantine bands[.][95]

In *La recherché de l'absolu* [*The Alkahest; or, in search of the absolute*] (1834) by Honoré de Balzac (1799–1850), Monsieur Balthazar Claës-Molina, Comte de Nourho, brings his family to a state of financial ruin in an alchemical quest, seeking to combust diamonds in order to understand how to synthesize them, and through this reverse engineering become a 'co-worker with Nature!'.[96] His apparatus resembles those used by Lavoisier and Davy, and – like Birmingham's statue of Priestley (see Chapter 2) – Balthazar's own intense observation is associated with the combustive gaze of the burning-glass:

> His eyes were gazing with horrible fixity at a pneumatic trough. The receiver of this instrument was covered with a lens made of double convex glasses .... [T]he machine ... was placed on a movable axle so as to keep the lens in a perpendicular direction to the rays of the sun. (pp. 211–12)

Relying upon the Sun to 'put our hands on the great secret' and 'enrich us all', both Balthazar and his valet Lemulquinier become obsessed with the availability of good weather to facilitate their research (pp. 213, 216). At a crucial moment of the novel, his daughter Marguerite disturbs Balthazar at work, in a reverential pose resembling Joseph Wright of Derby's *Alchymist* (1771), and the Sun's rays become an instrument of psychological rather than chemical analysis:

> The aspect of her father, half-kneeling beside the instrument, and receiving the full strength of the sunlight upon his head, ... his face contracted by the agonies of expectation, the strangeness of the objects that surrounded him, the obscurity of parts of the vast garret from which fantastic engines seemed about to spring, all contributed to startle Marguerite, who said to herself, in terror,— 'He is mad!' (p. 212)

The different perspective brought to light reveals not only Balthazar's obsession but also Marguerite's own tendency towards the irrational, as she is caught up in a moment of laboratorial reverie, with the scientific experiment and its instruments transfigured into forms of gothic terror. Eventually, Balthazar discovers a diamond apparently crystallized in the apparatus, although its origin is never explained, and some characters suspect one of the family's gems has been placed there to put him out of his misery. Nevertheless, the tragic irony of Balthazar's quest is that his life and social relations combust around him: like Victor Frankenstein's, his life serves partly as a stark warning of the risks of scientific ambition.

A Century Hence: Or, a Romance of 1941 (c.1841), by the American politician and author George Tucker (1775–1861), meanwhile, tells of an 'immense convex mirror of thirty feet diameter' at Greenwich which 'burns a diamond like a bit of paper in a candle'.[97] Like *The Alkahest*, this novel provides a cautionary tale to those caught in the zeal of scientific pursuit, since 'a young physician, in the ardor of experiment by which he hoped to extract a new metal from crystal, … had his hand burnt off'. In contrast to Balzac, however, Tucker finds black comedy in this situation, since 'as the injury was not attended with the usual consequences of burns, nor the same tediousness in healing, it has been recommended as an improvement in surgery' (p. 69).

Sun-treaders

While technologies of light were producing new knowledge in the hands of natural philosophers such as Davy, literary texts evince a cultural fascination with moments of solar energy encountered as a way of thinking through new ideas of the Sun's heat and light. The imaginary solar technologies and agencies involved are instruments of not only physical but also metaphysical, psychological or social transformation and/or discovery. Burning-glass experiments and their fictional analogues functioned as metaphors *and* metonyms of scientific discovery – as literal manifestations of enlightened progress bringing forth new knowledge and realizing Promethean ambitions. Yet these moments of radiant transfer and thermally induced discovery, sometimes figured as exchanges of visibility and invisibility, are paradoxically both revelatory and reductive.

Many accounts of Percy Shelley's youth refer to his avid interests in chemistry, natural philosophy and particularly optics, encouraged by the presence of Adam Walker (1730/31–1821) and James Lind (1736–1812) as tutors at Eton. Into his adult life, Shelley cherished the otherworldly spectacles generated by his solar microscope,[98] but at an early age his obsession with the Sun was expressed in violent fashion. The *Quarterly Review* suggested that, at Eton, Shelley was 'notorious for setting fire to old trees with burning glasses, no unmet emblem for a man, who perverts his ingenuity and knowledge to the attacking of all that is ancient and venerable in our civil and religious institutions', framing his prodigious pyromania as a Promethean act of rebellion.[99] His university friend and biographer Thomas Jefferson Hogg (1792–1862) found it 'altogether incredible' that a simple burning-glass could cause such destruction, but heard that the lens had been used to ignite gunpowder placed inside a tree's hollow. Hogg thought it a 'very trifling affair', but conceded, at least half-seriously, to the *Quarterly Review*'s claim that in scaring a schoolmistress and burning down woodland 'you might foresee the future opponent of superstition and tyranny'.[100]

Andrew Amos (1791–1860), Shelley's peer at Eton, believed that Adam Walker's lectures were responsible for encouraging his destructive nature, supplying him with 'the means of producing interesting and dazzling results requiring very little application of mind'. Shelley was (like Frankenstein) 'a youth carried away by an impetuous enthusiasm for producing and witnessing phenomena of nature', and particularly those involving combustion. By contrast, Shelley called Amos an '*Apurist*; ... one who did not appreciate properly the element of *fire*'.[101] What ideas Walker taught Shelley is not known precisely, but the first lecture of his *A System of Familiar Philosophy* (1799) is devoted to the Sun and its light. Possibly following Herschel, Walker thought the Sun's rays were generated in its atmosphere, in an 'ocean of fluid fire'. Unlike Herschel, however, Walker saw fire, light and electricity as aspects of the same elementary principle, and this was apparent in their interaction with human technologies: 'The Sun [is] the fountain-head of fire in our system; and ... his light, when it reaches the earth, is only that fire in a greatly rarified and mixed state; and, therefore, when condensed by lenses or mirrors, it becomes real fire again.'[102] Walker even speculated that the 'impulse of light' was responsible for diurnal motion and, again, this was suggested by the ability of concentrated light in a burning-glass to make molten gold turn on its axis (p. 4). The effect of focused light on opaque bodies, in melting and vitrifying, showed its affinity with fire and electricity (pp. 327, 328). Burning-mirrors and lenses were, therefore, crucial to Walker's conception of light. Shelley was also an avid reader of Davy, whose *Elements of Chemical Philosophy* (1812) he had owned since its publication, a book which made much of the potential of thermal optics in harnessing 'invisible rays'.[103]

The Sun fascinated Shelley throughout his poetic career, and can be found almost everywhere in his work. Soon after acquiring Davy's *Elements*, Shelley wrote the 'Sonnet: To a Balloon, Laden with *Knowledge*' (August 1812), in which the fire-balloon is a kind of 'Sun which ... / Shall dart like Truth where Falsehood yet has been' and sends 'A ray of courage to the oppressed' (ll. 13-14, 9; *Poems*, I, 239–40). *Queen Mab* (1813) is probably the first work in English to use the phrase 'solar power', although this seems to refer to the Sun's gravitational attraction, rather than the supply of photonic energy.[104] The Sun's role in powering the Earth is emphasized particularly in his two solar 'hymns'. His translation of the Homeric hymn 'To the Sun' (1818) acknowledges (anthropomorphically) how the Sun generates currents in the atmosphere (ll. 16-20; *Poems*, II, 342), while in 'Song of Apollo', written from the perspective of the solar deity himself, Shelley follows Walker in identifying the Sun as the source of all light: 'Whatever lamps on Earth or Heaven may shine / Are portions of one spirit; which is mine' (ll. 19-20, 23-24; *Poems*, III, 352). This statement is inserted between two Platonic explorations of his powers, the first of which asserts the Sun's role as bringer of truth: 'sunbeams are my shafts, with which I kill / Deceit' (ll. 13-14; *Poems*, III, 351).[105] The second acknowledges Apollo's creative and revelatory role as god of poetry: 'I am the eye with which the Universe / Beholds itself, and knows it is divine' (ll. 31-32; *Poems*, III, 352). Similarly, *A Defence of Poetry* asserts poetry's Promethean role in 'ascend[ing] to bring light and fire from those eternal regions', while in an extended metaphor on the power of the imagination, 'The Triumph of Life' (1822) describes how 'flashing rays' send the Sun's generative powers to Earth.[106] Given Shelley's preoccupation, it is no wonder that his public image as a writer was framed in solar terms. Thomas Lovell Beddoes (1803–49) called him 'A spirit of the sun, / An intellect a-blaze with heavenly thoughts'. For Robert Browning (1812–89), Shelley was the 'Sun-treader', 'a star to men' whose imagination could dwell upon the most sublime of things, a concept celebrated musically in the

disconcerting, dissonant *Sun-Treader* (1931) by the American composer
Carl Ruggles (1876–1971).[107]

Shelley's fullest exploration of the imaginative possibilities of solar energy is
in *Prometheus Unbound* (1819). Indeed, Prometheus's own creation is compared
to 'a beam / From sunrise' (I, 157–8; *Poems*, II, 487). Shelley grants his central
mythic figure a more conciliatory aspect than we might expect, and the solar
rays are not destructive, but 'life kindling' reflections of universal love (III.
iii.118; *Poems*, II, 593).[108] Radiant energies abound, none more so than in Act
IV when Panthea describes a ray of light projecting from the Spirit of the Earth:

> And from a star upon its forehead, shoot,
> Like swords of azure fire, or golden spears
> With tyrant-quelling myrtle overtwined,
> Embleming heaven and earth united now,
> Vast beams like spokes of some invisible wheel
> Which whirl as the orb whirls, swifter than thought,
> Filling the abyss with sun-like lightenings,
> And perpendicular now, and now transverse,
> Pierce the dark soil, and as they pierce and pass,
> Make bare the secrets of the earth's deep heart;
> Infinite mines of adamant and gold,
> Valueless stones, and unimagined gems[.] (IV, 270–81; *Poems*, II, 629–31)

This cosmic vision, which to some extent recalls Rev. 1.12-16, is brimming
with allegorical potential clustered around a technological metonym for
Enlightenment science, and loses its syntax as it tries to keep pace with this
thing 'swifter than thought' and able to project through all matter. From the
forehead, the seat of rationality, burst forth Platonic but weapon-like rays
of knowledge able to strike down tyrants.[109] These 'Vast beams' are light-
like in their immense speed, and the change in their direction suggests the
polarization (or 'transversality') of light discovered in the late seventeenth
century, but which had received renewed attention from Thomas Young.[110]
As 'spokes of some invisible wheel', they remind us of Helios' chariot, as
well as the 'flashing spokes' of the 'restless wheels of being' in Shelley's own
Queen Mab.[111] These 'sun-like lightenings' are also reminiscent of Davy's
statement that the 'refraction and effects of the solar beam offer an analogy
to the agencies of electricity' (*Elements*, p. 212), and of Walker's conflation of
electricity, light and heat.[112]

Most significantly, these beams have a revelatory power, penetrating the 'dark soil' to illuminate and liberate the 'secrets of the earth's deep heart', accessing Huttonian deep time and uncovering 'unimagined' materials, transforming the common trope that the Sun makes gems sparkle into a scientific action of discovery. Going a step further, 'The beams flash on' to bring to light not just geological but also archaeological knowledge: 'the melancholy ruins / Of cancelled cycles' (IV, 287–9; *Poems*, II, 631). The beams facilitate the production and restoration of knowledge, not the destruction of things. They might be sword- or spear-like but, in keeping with the optimistic, epithalamium-like tone of Act IV, these rays unite 'Heaven and Earth' and present and past, prefiguring the cosmic marriage of the Earth and the Moon in a later passage (see IV, 437–56; *Poems*, II, 641–2). These beams have been likened to electrochemical processes or X-rays,[113] although infrared imaging is the closest modern analogy. While it is unlikely there is a single source for Shelley's idea, the orb's abilities are certainly reminiscent of the laboratory burning-lens, which had generated many chemical discoveries by 'unit[ing]' heaven and earth. Yet these 'Vast beams' are not focused on a single point, but 'whirl[ing]', and this rotation physicalizes their revolutionary potential.

In contrast to Shelley's enthusiastic optical imaginary, for John Keats (1795–1821), the spectroscopic gaze would become an appropriate metonym for scientific reductionism. The mythical romance *Lamia* (1820) was written in the glorious summer of 1819, when Keats had 'fair Atmosphere to think in'.[114] Keats himself recognized the radiance that pervaded the poem, writing: 'I am certain there is that sort of fire in it which must take hold of people in some way'.[115] Keats's description of its readerly effects reflects the poem's preoccupation with literal combustion. The narrative is also structured by kinds of invisibility. At the start of the poem, we encounter Hermes as a Promethean figure of light and heat, 'bent warm on amorous theft'.[116] In an audacious pun, 'From high Olympus had he stolen light, / On this side of Jove's clouds, to escape the sight' of his lord (I, 9-10). His unbridled passion and constant radiance are bound together in a system of internal combustion, as 'celestial heat / Burnt from his winged heels to either ear' in contemplating the renowned Cretan nymph who has been rendered invisible by the serpent Lamia (I, 22-23). Both the lustful Hermes and the magical Lamia have powers over appearance, however, and in exchange for the ability to see the hidden nymph, the 'warm, flushed ... / ... burn[ing]' Hermes (I, 129–30) transforms Lamia into human form:

> Her eyes in torture fixed, and anguish drear,
> Hot, glazed, and wide, with lid-lashes all sear,
> Flashed phosphor and sharp sparks, without one cooling tear.
> The colours all inflamed throughout her train,
>
> ...
>
> So that, in moments few, she was undressed
> Of all her sapphires, greens, and amethyst,
> And rubious-argent; of all these bereft,
> Nothing but pain and ugliness were left. (I, 150–3, 161–4)

This violating encounter with the radiant messenger of light leaves Lamia scorched and 'undressed' of her precious metals, and much reduced of her 'dazzling hue[s]' (I, 47). Lamia's transformation, unique to Keats's version of the story, has been likened to an electrochemical reaction (or, less likely, an astronomical phenomenon),[117] but Hermes's feat is also reminiscent of a burning-glass vapourizing precious metals or minerals, separating them into constituent elements, or perhaps of ultraviolet rays decolouring silver chloride. Keats had taken two courses in chemistry while studying at Guy's where he would have learnt of not only the 'effects of light on the animal and vegetable economy' but also 'rays of light, which, though not discernible, produce chemical effects'.[118] In *Lamia*, light's transformative influence is mythologized as something of nature but which can violate and reduce it.

Hermes's destructive transformation of Lamia prefigures the philosopher Apollonius's rationalizing gaze which, at the end of the poem, can penetrate through Lamia's external self to the serpent beneath. The 'bald-head philosopher / ... fixed his eye' upon Lamia, and at this moment Lycius felt her hand grow 'hot, and all the pains / Of an unnatural heat shot to his heart' (II, 245–6, 252–3). 'Like a sharp spear' (II, 300), Apollonius's extramissive eye of reason burns its way to the truth, in a radiant act of both discovery and destruction. Introducing this final episode is the often-quoted passage which frames Lamia's unmasking as an allegory of scientific reductionism (of 'unweaving the rainbow').[119] Paradoxically, however, while Apollonius's eye generates heat to 'melt' Lamia 'into a shade', it is 'cold philosophy' against which the speaker rails for banishing the 'charms' of nature (II, 238, 229–30). Solar technologies often functioned as literal manifestations of enlightened progress bringing forth new knowledge. Yet, as we see in *Lamia*, these tools of analysis could be directed against the scientific project itself.

For all of the scientific, technological, literary and artistic understandings outlined in this chapter, perhaps the period's most compelling acknowledgement of the importance of solar power is an assertion of what the Sun's absence might mean. The massive eruption of Mount Tambora in Indonesia in April 1815 released volcanic aerosols which cooled the atmosphere for the next year or two, its effects eventually apparent in the northern hemisphere. It resulted in 1816 becoming the 'year without a summer', and brought harvest failures around the world. There also happened to be an increase in the number of sunspots, observed by North American astronomers but communicated across the Atlantic, and this produced some concern about the Sun's health. Byron's disturbing poem 'Darkness', written between 21 July and 25 August 1816 at Chillon, magnified these experiences and anxieties, imagining a time when the Sun's radiance is not impeded atmospherically but 'extinguish'd', and explores what the consequences of this might be. In so doing, the poem offers a prescient vision of the Earth ravaged for its resources, left a 'lump of death', as humanity pitifully attempts to compensate for this erasure of solar energy by burning anything it can, for a meagre moment of heat and light.[120] As civilization is forsaken for physical survival, the light of reason becomes a dim flicker. 'Darkness' conveys hauntingly what science can now enumerate and confirm: that the Sun supplies us with 17.3 terawatts of energy per year, and is responsible for the continuation of all organic life. Byron's gloomy vision is partially reprised by Thomas Campbell (1777–1844) in his apostrophe to the dying Sun in 'The Last Man' (1823), although the speaker there is comforted by the belief that in Heaven it will 'shine / In bliss unknown to beams of thine'.[121] As the nineteenth century progressed, the potential for the Sun to go out provoked increasing cultural anxiety. Yet, as the next chapter will explore, this period would also bring forth new ways of imagining how the Sun's energy could sustain human life and industry, possibilities that even in the twenty-first century have yet to be fully realized.

4

A time of 'solidified sunshine':
Victorian imaginaries of solar energy

Keeping the flame alive

'To Colebrooke Dale' (1785), a sonnet by Anna Seward, the 'Swan of Lichfield' (1742–1809), laments the industrial transformation of a Shropshire village, where organic sounds of the 'woods and vales' have been replaced by the 'ever-clanging forge'. Crucially, the speaker sees artificial 'dark-red gleams, / From umber'd fires on all thy hills', while the 'beams, / Solar and pure' are 'shroud[ed] with columns large / Of black sulphureous smoke' – a vision also captured in Philippe de Loutherberg's painting *Coalbrookdale by Night* (1801).[1] Seward's poem, therefore, chronicles the energy transition which facilitated the birth of modernity, as coal fires and their gaseous products emulate - but also dominate and obscure - the Sun. One might wonder, then, what place solar energy had in nineteenth-century Britain, deep in the 'Age of Coal', as the economist William Stanley Jevons called it.[2] Indeed, this period is the 'first time in history' when humanity (or the West, at least) 'ceased to be entirely dependent on the revenue of sunshine'.[3] In other ways, however, the nineteenth century defined what we now understand by, and how we continue to imagine, 'solar energy', and the Sun itself. The previous chapter showed how William Herschel pioneered research into the invisible powers of the Sun. Further into the century, advances in spectroscopy enabled a star's light to be 'read', divulging its chemical composition and suggesting its gaseous state.[4] As Balfour Stewart (1828–87) and Joseph Norman Lockyer (1836–1920) celebrated, 'Spectrum analysis has given ... a new fulcrum wherewith to move the great unknown by the lever of inquiry.'[5] Perhaps most pertinently from our twenty-first-century perspective, knowledge of infrared eventually led Joseph Fourier (1768–1830) and John Tyndall (1820–93) to discover that some atmospheric gases absorb large amounts of radiant heat and are responsible for the greenhouse effect.[6]

The term 'energy', pertaining to physical power (rather than the attribute or characteristic of a person), is a product of the nineteenth century, probably first used by Thomas Young (1773–1829) in his lecture 'On Collision' (published 1807).[7] The development of thermodynamics as a field within physics mid-century brought a new significance to the topic of heat, and most importantly the Sun's radiance, as an agent of physical processes. In contrast to the eighteenth-century physico-theologians who celebrated solar abundance, many nineteenth-century physicists suggested that the Sun could not possess an eternal store of energy, but must be depleting, on a timeline of lengthy but still worrying magnitude. These concerns are evident in scientific and literary treatments of the Sun until the 1920s, when it became all but certain that thermonuclear reactions powered the Sun and its stellar brethren.[8] Yet these anxieties were not universal,[9] and interest in utilizing solar power was in some ways reborn at this moment.

Moreover, while the Sun's physical powers seemed to diminish over the century, its cultural powers were sustained and renewed by various emulations, both technological and artistic. Solar physics was providing 'fresh fuel' for the 'poetic fire', wrote one commentator.[10] 'Sunbeam! what gift hath the world like thee?' asked Felicia Hemans (1793–1835), recalling Humphry Davy's appreciation of sunlight's physical *and* imaginative properties.[11] The belief in the primal force of the Sun as a generator of stories and images is almost omnipresent, but particularly prominent in the mythography of Max Müller (1832–1900), the art of J. M. W. Turner (1775–1851) and the criticism of John Ruskin (1819–1900).[12] The idea of using sunlight technologically shifted from combustion to image-making. Indeed, as Thomas De Quincey (1785–1859) wrote in *Suspiria de Profundis* (1845), the daguerreotype involved 'light getting under harness as a slave for man'.[13] The first published experiments involving the exposure of silver nitrate to light were by Thomas Wedgwood (1771–1805), but it was Joseph Nicéphore Niépce (1765–1833) who, in 1827, announced his development of the photographic process that would transform how the nineteenth century saw itself. Dubbing it 'héliographie' (Sun writing), Niépce's personification cemented the aesthetic importance of the Sun and its association with Apollo. Hercules Florence (1804–79) proposed the term 'photography' soon after, in 1834.[14] This light writing facilitated new knowledge of the Sun itself. No one had attempted to make a full sketch of an eclipse, for example, and relied upon memory for some of the details; but from the 1850s photography made full, but fuzzy, snapshots possible.[15] Astronomical photography became increasingly sophisticated and professionalized, and the popularity of books on the Sun in the late nineteenth century is no doubt partly attributable to their lavish photography of corona, which create their own sublime, chiaroscuro effects.[16] Photography could also capture parts of the spectrum invisible to the naked eye.[17] The Victorian era witnessed numerous attempts, therefore, to emulate and even surpass the Sun as a source of light and heat, at a time when the Sun's own powers were placed under anxious scrutiny. However, this chapter will explore the ways in which the direct capture of solar energy continued as a technological and imaginative endeavour, and played a part in influencing wider debates about the resources and social conditions of modernity, humanity's relationship with nature and the future of the solar system.

Browning in the Sun

As already noted in Chapter 2, the transition to coal affected the solar technologies that had burnt their way into the Enlightenment laboratory. Burning-mirrors and lenses had generally been replaced by improved coal-based furnaces and were no longer taken seriously as practical objects of research, although their continuing historical allure during the nineteenth century is evident in popular magazines and science books.[18] Their sublimity slipped into memory as the age of the industrial machine pressed on, and as other technologies, including the camera, came to occupy markets and imaginations. Yet the technical legacy of eighteenth-century thermal optics can be traced to innovations such as the polyzonal lens of the Marquis de Condorcet (1743–94), which was adapted by Augustin-Jean Fresnel (1788–1827) for use in lighthouses, and the hotbox and solar cooker of John Herschel (1792–1871).[19] Furthermore, while the authors of realist fiction generally concerned themselves with the shadows cast within coal-fuelled modernity, the burning-glass remained a rich but increasingly clichéd literary trope, with examples from Charles Dickens, George Eliot and Thomas Hardy, to name but a few.[20]

Most vividly, Charlotte Brontë's *Jane Eyre* (1847) shows us how the burning-glass's theological connotations continued to provide figurative nourishment. When Jane is sent to the 'evangelical, charitable establishment' of Lowood, Brontë (1816–55) deploys thermal optical metaphors to convey her heroine's plight.[21] As a cruel punishment for dropping her slate, Jane is humiliated by the headmaster, Mr Brocklehurst, in front of the whole school, and feels 'their eyes directed like burning-glasses against [her] scorched skin' (p. 76). Yet, the most intense experience, an overload of sentiment, occurs when she receives a sympathetic glance from one of the other pupils, whose smile was 'like a reflection from the aspect of an angel': 'She lifted her eyes. What a strange light inspired them! What an extraordinary sensation that ray sent through me! How the new feeling bore me up!' (pp. 77–8). When the girl herself, Helen Burns, is punished, Jane describes the 'fury' which 'had been burning in [her] soul all day', the 'tears, hot and large' continually 'scalding [her] cheek', and 'intolerable pain at the heart' (p. 86). Recalling the transverberation of St Teresa of Avila, and the optical similes in Anthony Horneck's meditations (see Chapter 1), Brontë deploys photothermic tropes of religious feeling to develop Jane's nascent interiority and to charge the evangelical school with moral hypocrisy.[22] The burning-glass was no longer an object of intense scientific interest and innovation, then, but continued to offer vibrant ways to conceptualize the self and the world.

Indeed, in his parleying 'With Bernard de Mandeville' (1887), Robert Browning (1812–89) reimagined the burning-glass as a kind of synecdoche of the Prometheanism that lay behind industrial modernity.[23] The collection *Parleyings* involves the speaker (ostensibly Browning himself) summoning and interrogating various historical figures, often in comparison with more contemporary voices. 'With Bernard de Mandeville', probably written autumn 1885,[24] starts as an exploration of the existence of evil, essentially pitting an 'optimistic' attitude Browning associates with the satirist and moral critic Mandeville (1670–1733) against the pessimism of Browning's dead friend Thomas Carlyle (1795–1881). This soon turns into a consideration of wider questions about the source and authority of man's knowledge.[25] Essentially, through a striking union of the mythic and technological, Browning revisits one of his dominant preoccupations: how one finds the 'infinite within the finite', as he famously wrote to Ruskin.[26] At the climax of the poem, Browning frames this difficulty by imagining humanity's early experiences, allegorized through the thoughts of a single primitive everyman:

'I solely crave that one of all the beams
Which do Sun's work in darkness, at my will
Should operate – …
To realize the energy which streams
Flooding the universe. …

If just one spark I drew, full evidence
Were mine of fire ineffably enthroned –
Sun's self made palpable to Man!'

Thus moaned
Man till Prometheus helped him, – as we learn, –
Offered an artifice whereby he drew

Sun's rays into a focus, – plain and true,
The very Sun in little: made fire burn
And henceforth do Man service – glass-conglobed
Though to a pin-point circle –[27]

In Browning's rationalized rewriting of the Prometheus myth, perhaps inspired by Lord Byron and Percy Shelley (see Chapter 3), the uncanny burning-glass as solar medium becomes a compelling symbol of the mind's eye, and of humanity's control over the natural world. This proxy for the Sun seems to generate a special kind of the technological sublime that, like the refraction of disparate sunbeams, fuses the celestial and the human within a microcosmic, tangible form.[28] This episode has been interpreted as a Christian allegory, with the Sun functioning as a metaphor for 'God's absolute being', and the act of bringing 'the very Sun in little' down to Earth to increase Man's knowledge hinting at Christ's Incarnation.[29] The climax of the poem proclaims the value of the 'little' (and of the external world) in enabling the human mind to 'infer [God's] immensity' (l. 317).[30] Given access to the transcendental burning-glass, Man can recreate this divine love repeatedly, as long as he keeps himself within the light.

Browning's later writings often explore the dialogue between the primitive and the modern, and can be viewed as part of a wider cultural exploration of myth that inaugurated modern anthropology, which included the work of Max Müller and Edward Burnett Tylor (1832–1917).[31] Müller, whom Browning knew personally, accorded the Sun a special place in his mythography. Almost all myths could be explained as an expression of devotion to our sovereign star, since 'it is generally, if not always, the sun or the sky which forms the bridge from the visible to the invisible, from nature to nature's God'. Müller argues,

> Not even the most recent scientific discoveries described in Tyndall's genuine eloquence, which teach us how we live, and move, and have our being in the sun ... – gives us any idea of what this source of light and life ... was to the awakening consciousness of mankind. ... man looked up to the sun, yearning for the response of a soul, and though that response never came, ... he never doubted that the invisible was there, and that, ... where he could neither grasp nor comprehend, he might still shut his eyes and trust, fall down and worship.[32]

This is probably a reference to Tyndall's *Six Lectures on Light* (1873), but, as Gillian Beer notes, Müller inflects the Sun with a quasi-divine status.[33] For Müller, solar spectroscopy and physics have extended humanity's senses and understanding, but the ultimate mystery of the Sun continues, and in that experience we are aligned with our primeval ancestors. In the parleying 'With Bernard de Mandeville', however, Browning follows the opposite analogical process, looking to the past to try to understand the present. This should not come as much of a surprise: despite the often archaic or timeless settings of his works, recent studies of Browning have attempted to suggest their engagement with cutting-edge issues.[34] The Promethean myth, which imagines the origins of technology (and in the form of the burning-glass gives a kind of Industrial Revolution in itself), allows Browning to explore humanity's relationship with nature, at a time when reliance upon but also threat to the environment is at an unprecedented level. Moreover, despite the everyman's in some ways simple ambition to start a fire, in his desire to 'realize' (perhaps in the senses of 'convert', 'bring to fruition' and 'understand') the 'energy which streams / Flooding the universe', his discourse is reminiscent of scientists and popular science writers who, as we will see, lamented the impending coal shortage, and the constant waste of the Sun's immense radiation. Unlike them, however, Browning writes a poem which does not mourn the 'death of the Sun' but is about utilizing its power, voicing a kind of pragmatic consolation that humanity will make good use of even 'one spark' – an optimism about the Sun as energy source shared by contemporaneous engineers, as the chapter will later explore.

Sun from Newcastle

In June 1768, Horace Walpole (1717–97) wrote to a friend to complain of the *'bad summer'* they were enduring and joked that 'the best sun we have, is made of Newcastle coal, and I am determined never to reckon upon any other'.[35] Walpole inadvertently hit upon a truth that would not be fully understood until the next century, observable in Davy's *Elements of Chemical Philosophy* (1812), which brought together various discoveries regarding photosynthesis.[36] One of the crucial ways in which the Victorians influenced more recent solar energy discourse is illustrated by the apocryphal story that George Stephenson (1781–1848), the renowned engineer, once set his friend William Buckland (1784–1856) this conundrum:

> 'Can you tell me what is the power that is driving that train?' ... 'I suppose it is one of your big engines.' 'But what drives the engine?' 'Oh, very likely a canny Newcastle driver.' 'What do you say to the light of the sun?' 'How can that be?' ... 'it is light bottled up in the earth for tens of thousands of years, – light, absorbed by plants and vegetables, being necessary for the condensation of carbon during the process of their growth, ... and now, after being buried in the earth for long ages in fields of coal, that latent light is again brought forth and liberated, made to work, as in that locomotive, for great human purposes.'

For Stephenson's biographer, Samuel Smiles (1812–1904), the great engineer's words were 'like a flash of light [which] illuminated in an instant an entire field of science'.[37] Given that Buckland was an eminent geologist, it is hard to believe he did not grasp what Stephenson meant in the first place.[38] Nevertheless, this exchange highlights how the discourse of solar sustainability was in the nineteenth century co-opted by the fossil-fuelled industrial West, as it 'liberated' solar energy that had been chemically stored for millennia (the 'solidified sunshine of the world's youth', as one anonymous article put it) for the benefit of 'great human' progress.[39] There were often natural theological aspects to this argument – that divine providence had seen fit to furnish mankind with this source and that nature never lets anything go to waste – but, as Allen MacDuffie explores, they would come up against the problems of finitude and thermodynamics later in the century.[40] In particular, it was realized how 'far from economically' the Sun's energy, via wood and coal, was 'warming our houses' and 'driving our engines'.[41] On the one hand, the idea that coal was 'natural' seemed to justify its continued usage despite the realization that this resource was polluting, finite and 'being used up at an alarmingly rapid rate'. On the other, that coal was ultimately solar in origin invited speculation that more direct and efficient, cleaner and less wasteful ways of utilizing the Sun's energy would be preferable, and might be possible.[42]

Few of the works of thermal optical imaginary traced in the previous two chapters make reference to the great shift to coal consumption across Britain (Arbuthnot's *Humble Petition* being a rare exception), but in the nineteenth century solar technologies were explored as ways of understanding this hydrocarbon transition. The invention of the steam engine in 1712 by Thomas Newcomen (1664–1729), and improved by James Watt (1736–1819) in 1763, heralded an energy shift responsible for the Industrial Revolution. By the start of the nineteenth century, Britain was a fully fledged fossil economy. The ultimate expression of the nation's reliance upon coal was perhaps William Stanley Jevons's *The Coal Question* (1865), although anxieties about its finitude can be traced to at least 1789.[43] Jevons (1835–82), drawing upon private correspondence with Tyndall as evidence, warned of the impossibility of finding a substitute power source for the nation's continued industrial dominance (*The Coal Question*, p. xvi). In Jevons's opinion, coal's solar origins 'entitle[d] it to be considered the best natural source of motive power' (p. 164). This store of force 'collected from the sunbeams for us' chemically and 'made to work for human purposes' is 'like a spring, wound up during geological ages for us to let down' (pp. 165, 164). This analogy, and the anthropocentric assumption that this energy is for 'us', emphasizes the logic and naturalness of the resource's usage. Jevons acknowledged that more plentiful stores of energy were in existence, but that the means to harness them had not been identified: 'The sun annually showers down upon us about a thousand times as much heat-power as is contained in all the coal we raise annually; yet that thousandth part, being under perfect control, is a sufficient basis of all our economy and progress' (pp. 143–4). Discovering the ability to control this superior source, however, would be to the detriment of Britain within the global capitalist economy, given its climatic disadvantage: 'Some day the sunbeams may be collected, or ... some source of force now unknown may be detected. But such a discovery would simply destroy our peculiar industrial supremacy' (p. 168). A few years later, Jevons believed that such a discovery was close at hand, when he learnt of John Ericsson's 'solar engine', which was intended to 'supply a new fuel in the place of coal, and a new motive power instead of steam'. Even though he was somewhat sceptical that an energy shift would happen any time soon, he still worried that 'the seats of industry will be removed to the sunny parts of the earth. In Manchester at any rate ... we are likely to find coal a source of sunlight [rather] than sunlight a competitor of coal.'[44]

Once coal was burnt to generate electric lighting, its ultimate solar surrogacy was complete. Promoting Thomas Edison's light bulb, his collaborator Francis Robbins Upton (1852–1921) begins an article by correcting the 'mistaken idea ... that this new light was intended to be a rival of the sun', instead 'it really is ... a rival of gas'. By the end of the piece, however, the author changes his mind, and the solar sublimity of the bulb is confirmed:

> The latent force accumulated during the primeval days ... is converted, after passing in the steam-engine through the phases of chemical, molecular and mechanical force, into electricity, which only waits the touch of the inventor's genius to flash out into a million domestic suns to illuminate a myriad homes.[45]

The arc light, meanwhile, coming from the discharge of electric current between two carbon electrodes, was so intense that – like the real Sun – it could not be looked at directly.[46] The ultimate expression of these solar-electric dreams was probably the proposal by the architect Jules Bourdais (1835–1915) to erect a 360-metre-tall 'Tour Soleil' for the 1889 Paris Exposition. At the topographical centre of the city, a huge arc light and reflector would be aimed to illuminate everything within a 3.5-mile radius. The project was overlooked, however, in favour of the designs of Gustave Eiffel (1832–1923).[47]

The solar–coal symbolism also reflected back onto vegetable life and the process of photosynthesis, since the 'solar beam is the agent which tears the atoms asunder, setting the oxygen free, and allowing the carbon to aggregate in woody fibre'.[48] Even John Ruskin, often critical of the sciences,[49] was clearly influenced by new conceptions of solar energy and its chemical operations upon the Earth. In *Modern Painters* (1843), photosynthesis seems to inform his description of the trees on the Alban hills in Italy. 'Like the curtains of God's tabernacle', Ruskin sees 'every separate leaf quivering with buoyant and burning life; each, as it turned to reflect or to transmit the sunbeam, first a torch and then an emerald'.[50] The beauty of each leaf is a product of the flow of energy, its outward activity symbolic of a more powerful one within. By 'reflect[ing]' the light, the leaf acts like a kind of concave mirror, 'burning' with life; in 'transmit[ting]' the sunbeam, it absorbs certain frequencies of the visible spectrum, and the resultant 'emerald' is a vibrant expression of the photosynthetic process, with plant life continuing to mediate the Sun's material and imaginative nourishment.[51]

Although fossil fuels enjoyed a special position as 'bottled' light, it became widely recognized that almost all activities on Earth are, as Tyndall puts it, 'special forms of solar power – the moulds into which his strength is temporarily poured'. Humans, 'in a purely mechanical sense', are therefore 'the children of the sun'.[52] For Ruskin, again hinting at the importance of photosynthesis, Tyndall's concept had 'thrown foolish persons into atheism, [but] gives us the arithmetical and measurable assurance that men vitally active are living sunshine, having the roots of their souls set in sunlight, as the roots of a tree are in the earth'.[53] As MacDuffie has explored, within Tyndall's solar rhetoric, 'industrial activity and resource exploitation seem not only thoroughly natural but inevitable expressions of solar agency', influencing attitudes to fossil fuel for generations to come.[54] Whether this resource was natural or not, however, many would follow Jevons in dreading that humanity was 'rapidly squandering' nature's 'capital' on Earth.[55] Moreover, it was thought that the '*noxious gases* from manufactories' were, somewhat ironically, hampering the growth of plant life, through causing both '*deficiency of light*' and the '*dryness of the atmosphere*'.[56] Even worse than this, though, the original source was itself feared to be running out.

The great storehouse

Over the eighteenth century, there had been increasing thought and debate regarding the nature of the Sun and its rays, culminating in the groundbreaking work of William Herschel in the 1790s (see Chapter 3). In Query 11 to the *Opticks*, Newton had asked: 'Are not the Sun and fix'd Stars great Earths vehemently hot, whose heat is conserved by the greatness of the Bodies[?]'[57] What Newton did not speculate upon, however, was an issue crucial to those who maintained a material theory of light: that the emission of light particles would involve a constant loss of mass that would eventually leave the Sun, in the words of Lemuel Gulliver, 'wholly consumed and annihilated'.[58] Priestley and many of his contemporaries, however, followed the estimate by John Michell (1724–93) that the Sun would lose less than two grains per day.[59] Others found ingenious ways to explain that the Sun's energy production is unrelated to material expenditure, and therefore eternal.[60] But crucial developments in physics mid-century meant that the question of solar energy's source became difficult to ignore.

While the scientific concept of 'energy' has been traced to the late eighteenth century, it was during the 1840s that interest in steam engines and the thermal effects of electromagnetic phenomena led several scientists to conceive independently that an indestructible force was needed to supply natural processes.[61] Julius Robert Mayer (1814–78), James Prescott Joule (1818–89), Hermann von Helmholtz (1821–94), Rudolf Clausius (1822–88) and William Thomson (1824–1907) (later Lord Kelvin) all made major contributions to what in 1854 Thomson called the study of 'thermodynamics', laws concerning energy's conservation and dissipation (or, as Clausius termed it in 1865, 'entropy'). Based on the idea of a 'mechanical equivalent of heat', 'energy' could be used to denote the measurable amount of work done by anything which involved motive power, from engines to animal movement.[62] As Ted Underwood has shown, the solar origin of terrestrial labour became a predominant theme of such discussions.[63] Significantly, thermodynamics changed not only the physical sciences but also the cultural understandings.[64]

Particularly following Claude Pouillet's measurement (using a pyrheliometer) of the 'solar constant' (the amount of energy received from the Sun per unit area, in a given time), it was realized that the Sun's radiant energy had to be an immense resource. Pouillet (1790–1868) made a startling calculation: that the total annual influx was equal to that which would melt a global blanket of ice of 14 metres thickness. An external source was the most popular suggestion initially: that the Sun's gravitation drew a huge number of meteors or asteroids into itself constantly, replenishing its energy kinetically. John James Waterston (1811–83) and Mayer each struggled to publish their papers on this topic in the 1840s, so Thomson's development of Waterston's ideas, 'On the Mechanical Energies of the Solar System' (1854), became the most famous articulation of the meteoric theory.[65] Unlike Mayer, Thomson did not believe that the amount of heat the Sun lost would be completely replenished by the meteors, leaving the Sun to eventually run down. Thomson's paper dismissed the other two possibilities: that of internal chemical reactions and that the Sun was an originally heated body gradually radiating away its heat.[66] Meanwhile, Helmholtz, who had reviewed Waterston's paper, came up with an alternative gravitational theory: that the Sun was slowly contracting in on itself, releasing light and heat, but 'sufficient to maintain for an additional 17,000,000 of years the same intensity as that which is now the source of all terrestrial life'.[67]

Thomson abandoned the meteoric theory, and in his article 'On the Age of the Sun's Heat' (1862) in the popular journal *Macmillan's Magazine,* followed Helmholtz in arguing that the Sun 'originated in a coalition of smaller bodies, falling together by mutual gravitation', generating (according to the first law of thermodynamics) 'an exact equivalent of heat for the motion lost in collision'.[68] For the rate of solar combustion, Thomson made direct comparison to the burning of coal: 'The sun radiates heat, from every square foot of his surface, at only about 7,000 horse power. Coal, burning at a rate of a little less than a pound per two seconds, would generate the same amount' (p. 392). Following the second 'great' law, mechanical energy has a 'universal tendency' to dissipate, resulting in the 'diffusion of heat, ... and exhaustion of potential energy', even though empirically 'we do not know that he [the Sun] is losing heat at all' (pp. 388, 390). Thomson was tonally less optimistic than Helmholtz about the future: 'Inhabitants of the earth cannot continue to enjoy the light and heat essential to their life, for many million years longer, unless sources now unknown to us are prepared in the great storehouse of creation' (p. 393). As Gillian Beer has shown, an anxiety about the shrinking and cooling Sun resulting from Thomson's article generated 'much imaginative thought and production', and the enthusiastic reception given to Max Müller's work on solar mythology was perhaps in part a symptom of such displaced fears. While moments of shared interest in the Sun are manifest in various disciplines of knowledge in late-Victorian culture (even if the methodologies and precise subjects were very different), Beer identifies that there were also forms of discontinuity, between a system of physics that insisted upon increasing disorder and entropy, and Darwinian theories that foresaw development and complexity. Humanity was thought to have the potential to improve physically and intellectually, but increasingly difficult environmental conditions in the future might drive them to extinction.[69] Moreover, the forecasts for the Sun's duration were shorter than the 100 million years required for natural selection to have shaped life on Earth, as Thomson hinted provocatively ('On the Age of the Sun's Heat', pp. 391–2). Meanwhile, the prospect of heat death was also a compelling trope for thinking through economic and social concerns about energy expenditure and waste.[70] From a cosmic point of view, the Sun's energy was far from renewable.

While anxieties about the Sun's lifespan were tangible across Victorian culture, it would be erroneous to see them as universal. Despite conceding that 'our own Sun is not the youngest and brightest of all the suns and stars', the French astronomer Jules Janssen sought to reassure his readers that 'it has the potential to satisfy the most ambitious dreams of mankind'. Sir Robert Stawell Ball (1840–1913), in his popular book on the Sun (1893), asserted the validity of the contraction theory, but argued that 'the gain of energy by the molecules through gravitation towards the Sun compensates for their losses in virtue of radiation'. More extravagantly, Charles William Siemens (1823–83) could not entertain the idea that the Sun was so wasteful, with only a millionth of its energy captured by the planets (and the rest lost), and so proposed that space is full of gases in a system of recuperation 'somewhat analogous' to the workings of a 'regenerative gas furnace'.[71] There were also ambitious engineers for whom the measurements of solar influx were encouraging, not worrying, as they sought to find ways to capture this immense energy otherwise lost.

Engines of light

William Thomson outlined a dynamical theory of radiant heat and light in which the Sun's energy could be transformed. He conjectures, for instance, that 'mechanical effect of the statical kind might be produced from solar radiant heat, by using it as the source of heat in a thermo-dynamic engine', since Pouillet had estimated that 1 horsepower could be produced by an engine receiving 1800 square feet of rays.[72] It did not take long for some industrialist inventors to attempt this practically, with optical designs anticipating solar collectors of the past century or so, but drawing upon earlier concepts. Most notably, in the 1860s the French mathematician Augustin Mouchot (1825–1912), who had studied the thermal optics of Buffon and others vociferously,[73] developed a solar steam engine made of a cone-shaped mirror surrounding a copper cylinder filled with water. It was successful enough to receive national government sponsorship for a time, the 'conquest of the sun' being part of the Third Republic's modernizing and colonialist propaganda.[74] This subsidy evaporated, however, when war was declared on Prussia in July 1870. Fortunately, Mouchot then secured funds from the regional government of the wine-producing district of Indre-et-Loire to build a solar-powered distillery. Following this, Mouchot was paid to explore the potential for solar cooking, distilling, and pumping in Algeria, and the colonial application of his devices became an important part of his solar rhetoric.[75] At the 1878 Universal Exposition in Paris, Mouchot represented the French colony with a huge device: a mirror over 13 feet in diameter, with a boiler more than 6 feet long and containing 16 gallons of water (Figure 13). In the sunshine of 2 September 1878, it managed to boil the water within half an hour and produce a pressure of 6 atmospheres.[76] Despite this relative success, Mouchot returned to teaching mathematics, with his solar research continued by his assistant Abel Pifre (1852–1928), who built several solar-powered motors, including one demonstrated in the Gardens of Tuileries in Paris, which drove a press and printed 500 copies of his *Journal Soleil*. The French government sponsored further tests of the machines, but decided that they were not efficient enough to receive subsequent investment.[77]

Figure 13 M. Férat, 'Le Soleil a L'Exposition (Trocadéro)', *Le Monde Illustré*, 22: 1125 (19 October 1878), p. 252.

For a time, Mouchot's machines drew comment on both sides of the channel, inspiring authors to consider sustainable futures characterized by new kinds of social and geopolitical order, and strengthening arguments about current energy wastage which had 'excited the displeasure' of economists like Jevons. An anonymous essay in *The British Quarterly Review*, for example, frames a story reporting Mouchot's work with a discussion of energy supplies:

> Is it not distressing to know that the beams which play so unprofitably, in some respects, on many parts of our earth, might, if properly impounded and harnessed to cunningly-constructed machines, be compelled to serve mankind in a very useful and lucrative capacity?[78]

The article reports that Pouillet's prediction had been almost proven by Mouchot, with 1 square metre of solar radiation 'communicat[ing] as much heat every minute as would suffice to raise at least one litre (1.76 pints) of ice-cold water to the boiling point. In other words, ... nearly equal to the theoretical duty of a single horse-power steam engine' (p. 413). This was in Paris, and the article acknowledges that there are other places around the globe where 'the Lord of Day is known to stalk in burning splendour' and even greater achievements could be made (p. 413). The verb 'stalk' is used here primarily in the sense of 'march proudly through', but also hints at a kind of vegetative abundance associated with the Sun's power.[79]

Aesthetically though, Mouchot's work left much to be desired, according to some. An essay in *All the Year Round* (published at the time of Charles Dickens, Jr.'s stewardship) describes Mouchot's solar engine in a half-sardonic tone, as 'a queer-looking object – a silvered reflector, shaped like a huge lamp-shade turned upside down', far from a display of technological sublimity.[80] Despite its smaller scale, Mouchot's solar cooking stove (a glass vessel in the focus of one of his reflectors) was 'a greater wonder than his steam-engine – the wonder being that people never thought of making it before', with its usefulness foreseen especially 'over the African deserts, or the steppes of Central Asia' (p. 491).[81] The author also imagines, mock-heroically, a time when the solar stove is used universally across Britain as a domestic appliance, installed at the top of the house: 'Think what a boon that would be, ye whose genteel ten-roomed house is never free from a pervading smell of soup and fish! ... pray that M. Mouch[o]t may soon come to the rescue' (p. 493). Further ahead, the essay contemplates a time when the coal deposits are all but gone, and wonders whether 'we [the British] shall have to go and live within the tropics, [giving] up the countries which for many centuries have been the chief seats of civilisation'. Caricaturing Jevons's concerns about Britain's post-fossil fuel economy, the article contemplates a dystopian energy future of radically shifted geopolitics in which 'summer travellers would, with fear and trembling, make a tour in England or France, well armed with Mouch[o]t's solar cooking-stoves or the latest improvements upon them, and condemned to cold meat, and, worse, to cold limbs, and cold in the head, if the sun should decline to show himself for a day or two'. The author then changes tone, confessing: 'Seriously, for our own sakes as well as for posterity, it might be worth our while to give M. Mouch[o]t's machines a trial' (p. 492).

By the time the article appeared in *All Year Round*, Mouchot had a transatlantic rival. The Swedish American industrialist John Ericsson (1803–89), who had previously published several papers on the measurement and source of solar energy, displayed an immense 'solar engine' as part of the 1876 centennial exhibition: the first World's Fair, held in Philadelphia.[82] The engine used concave mirrors to concentrate a 'pencil of solar rays' covering 35 square feet upon a steam generator, but, for commercial reasons, Ericsson was secretive about the precise technology employed.[83] He envisaged that 'those regions of the earth which suffer from an excess of solar heat will ultimately derive benefits resulting from an unlimited command of motive power', but did not recommend its utilization elsewhere, until a means of storing the radiant energy had been devised (pp. 561, 563). Ericsson acknowledged the work of Mouchot but was dismissive of his achievements, noting that John Herschel had developed a solar cooker many years previously (something not realized by the author in *All the Year Round*), and that a hundred of Mouchot's generators would be needed for even 'very moderate power' (pp. 564, 569).[84]

While Mouchot's technology was (for a time) co-opted by the state as a way of projecting colonial power, in Émile Zola's hands the possibility of capturing solar energy became part of a socialist vision. Zola (1840–1902) had attended the 1878 Universal Exposition many times, and Mouchot's devices may have inspired the techno-utopian ambitions of the scientist Jordan in *Travail* (1901).[85] The novel's last book follows the community of Beauclair, and as part of the area's utopian re-fashioning along the lines of Charles Fourier's self-sufficient, socialist '*phalanstères*', the seventy-four-year-old Jordan is seeking to reduce the entropy of its machines.[86] Fearing a time 'when coal would be exhausted', Jordan's main obsession is to find an alternative means of energy, and particularly one which would make the 'sun himself the universal motor', drawing 'directly from his rays that calorific power contained in coal'. He also seeks to conquer the problem of storing 'solar heat'.[87] Unlike Balzac's chemist Balthazar in *La recherché de l'absolu* (1834) (see Chapter 3), however, the similarly fanatical Jordan's projects are unambiguously (but no less miraculously) successful. One design is able to 'directly transfor[m] the caloric energy contained in coal into electrical energy without passing through mechanical energy' (p. 578). Another 'render[s] the reservoirs' of solar energy 'impermeable', preventing any loss (p. 580). Zola's utopian climax, in which the 'gleam of its rays of glory sparkled the roofs of Beauclair triumphant' (p. 603), therefore rests upon a violation of the second law of thermodynamics, which ultimately undermines its socialist potency.[88] Yet the turn of the century produced an even more radical vision of solar power engineering which foregrounded questions of resource and consumption in vivid, violent fashion.

Heat-Rays and heliographs

Long before William Herschel transformed our knowledge of sunlight, he had considered the extraterrestrial use of solar power. Sometime in the late 1770s, he wrote down his thoughts concerning the possibility of life on the Moon. He speculated that 'Lunarian' intelligences would live in circular buildings: 'For in that Shape … one half will have the direct and the other half the reflected light of the Sun.'[89] More violently, in Washington Irving's *A History of New York* (1809), the pseudonymous author 'Diedrich Knickerbocker' warns of Earth's colonization by the people of the Moon. This anti-imperialist satire imagines the Lunarians 'as superior to us in knowledge, and consequently in power, as the Europeans were to the Indians, when they first discovered them', 'riding on Hypogriffs, defended with impenetrable armour – armed with concentrated sun beams'.[90] The fantasy – nay, anxiety – that advanced alien civilizations can utilize solar power more efficiently and effectively, and to devastating effect, remains an enduring trope of science fiction. Although the immediate influence of solar energy technology upon the literary imagination is limited and diffuse during the nineteenth century, this period was important in laying the foundations of the SF genre, and ideas of solar energy were crucial to that development. This paradigmatic role in the imagining of solar futures is nowhere more apparent than in H. G. Wells's science fiction, which exploited the consequences of a cooling Sun, particularly *The Time Machine* (1895) and *The War of the Worlds* (1897). The scientific romances of Wells (1866–1946) continue to be a touchstone for SF writers in so many ways, including how they imagine advanced weaponry and cosmic apocalypse, both of which can be traced to the nineteenth-century solar imaginary, if not earlier.

Wells's solar ideas did not come out of nowhere. During the 1890s, Wells reviewed several works concerning our nearest star. Hitherto unexplored, the works and their reviews provide new insight into Wells's solar imagination while crafting his greatest works of SF; they also underline the important but complex relationship between popular science and SF at a significant moment in their joint development. The first book under Wells's consideration was *The Dawn of Astronomy* (1894) by the physicist Joseph Norman Lockyer. His privately established Solar Physics Observatory in South Kensington was just over the road from the Normal School of Science that Wells had attended in the 1880s, and Lockyer lectured at the school from time to time.[91] Lockyer's most significant contribution to solar science was discovering helium through spectroscopic analysis of the chromosphere. His latest book, however, examined sun worship among ancient civilizations in order to trace the origins of astronomy. Lockyer painstakingly explored how the solar temples of Ancient Egypt, Babylon and elsewhere were oriented to observe the Sun and be illuminated by its rays in patterns that heightened their rituals in praise of Amen-Rã, Chnemu and other solar deities. The book is abundant, therefore, with descriptions of sunlight beamed through apertures onto statues, or reflected in jewels, in order to stage the 'whole solar drama' between gods of light and dark.[92]

Wells reviewed Sir Robert Stawell Ball's popular physics book *The Story of the Sun* (1893) just a few weeks after Lockyer's. Ball, appointed Lowndean Professor of Astronomy and Geometry at Cambridge in 1892, had produced a work of popular physics using the latest astronomical research. Its topics include everything one might expect, with chapters on the Sun's mass, distance from the Earth, corona and so forth. What sets it apart from similar books appearing at this time – such as *The Sun* (1881) by Charles Augustus Young (1834–1908), Professor of Astronomy at Princeton – is the use of analogy to enhance the reader's comprehension, combined with a sense of narrative regarding not only the Sun's 'life' but also the transfers of energy which produce its phenomena. For example, Ball explains that 'Sun-beams hasten towards us with their marvellous story at a speed one hundred times as great as that with which the cable conveys [signals] beneath the waters of the Atlantic'. The role of the scientist, or science writer, is to narrate the 'story' the sunbeam contains within itself. Part of this tale, of course, is where these sunbeams end up, with 'almost all the power that is displayed in exercise around us' due to their action, while sunbeams of 'untold ages ago' result in the heat produced by the 'union of coal with oxygen'. Ultimately, like Lockyer, Ball presents the 'story of the sun' as also the story of humanity, but one determined by the 'mechanical equivalent of heat'.[93] Ball is also innovative in his addresses to the audience, asserting that 'the eye of the reader, as he follows these lines, passes along the text with a velocity which may be regarded as the million-millionth part of the speed with which the message from the words is transmitted to this pupil' (pp. 62–3). Human physiology pales in comparison to the sublime powers of light. Nevertheless, Ball recognizes the importance of handling carefully the reader's experience, seeking to enhance scientific understanding through narrativizing one of the greatest stories in nature.

Wells admired the plain style and 'inhumanity and serene vastness' of Ball's subject. He found the idea of electromagnetic tides brushing by 'our little eddy of planets', unsettling our compasses, making solar storms, then passing on to 'the illimitable beyond', 'so powerful and beautiful as to well-nigh justify that hackneyed phrase, "the poetry of science"'. What Wells identified as 'especially interesting', though, was the 'speculation upon the physical condition of the sun, on its beginning, and on the radiation away of its energy, and its slow shrinkage as time passes away'.[94] Ball's book presumably fuelled Wells's vision of the future Earth in *The Time Machine*, when the Traveller observes 'the sun grow larger and duller in the westward sky, and the life of the old earth ebb away'.[95] In his essay 'The Discovery of the Future' (1902), meanwhile, Wells confirmed he thought it was with

> reasonable certainty that this sun of ours must radiate itself toward extinction; ... it will grow cooler and cooler, and ... some day this earth of ours, tideless and slow moving, will be dead and frozen, ... There surely man must end. That of all such nightmares is the most insistently convincing.[96]

The Time Machine was published first in serial form in *The New Review*, between January and May 1895. In April of that year, Wells wrote a review of Isaac W. Heysinger's *The Source and Mode of Solar Energy throughout the Universe* (1895). Heysinger (b.1842) was an American inventor and amateur scientist, seemingly well read in the scientific authorities Helmholtz and Thomson, but who had composed his own theory regarding the 'grandest and most important question of all physics': the source of the Sun's power. Drawing analogies with electrical induction and electrolysis, Heysinger proposed that planetary rotation produces 'electrospheres' which then 'discharge their tremendous currents directly into the sun', maintaining harmony and balance within the closed (solar) system.[97] Wells is brilliantly ironic in his dismissal of Heysinger's chemical thesis:

> It is simply wonderful what insight unhampered by excessive knowledge can manage. The sun is a kathode, and the planets anodes, and so the attenuated water vapour (you know) between us and the sun is electrolized, and the hydrogen goes to the sun and the oxygen comes to the earth, and the sun is luminous because the positive pole of the arc light is. And what more do you want?

Wells tries to be fair to Heysinger, though, praising him for writing 'neatly and sometimes strongly', but finding the problem lying in the author having read 'nothing but a loose type of scientific book' and mimicking their style and 'trick of seeming to reason'.[98]

Despite the absurdity of Heysinger's central concept, there remain elements within the book that closely resemble ideas and images that would feature in *The War of the Worlds*, reminding us that chains of influence are not always fuelled by admiration. Heysinger includes, for example, a vivid description of the Sun 'sink[ing] into eternal frigidity'; a discussion of arc lamps and the 'numerous and powerful', 'obscure and invisible' heat rays they produce; and, perhaps most significantly, the speculation that 'each [planet] has had, or will have, its stage in which life is possible, and these planets may be like human habitations, in which whole races at times migrate from one home to another' (*The Source and Mode of Solar Energy*, pp. 21, 85, 129). Of course, Heysinger's own book is a tissue of citations, but it brings together many ingredients which would also coalesce (but in radically different ways) in Wells's story of alien invasion.

The 'nightmare' of the dying Sun is the ultimate premise of *The War of the Worlds*. In its opening chapter, the nameless narrator speculates as to why the Martians invaded:

The secular cooling ... which to us is still incredibly remote, has become a present-day problem for the inhabitants of Mars. ... To carry warfare sunward is, indeed, their only escape from the destruction that, generation after generation, creeps upon them.[99]

Although the apparent cooling of the solar system is implied to be the ultimate reason for Martian invasion, the solar energy context of the novel goes much further than this. First, images of the Sun upon the Earth (shining through windows, glinting upon tripods, and so on) form part of the story's visual symbolism, and Wells's techniques here in flashing light across the novel are well documented.[100] In part, the imagery serves to convey that the Martians have come to dwell on a planet where the Sun's rays still shine powerfully. Some of the novel's preoccupation with light is also no doubt related to the 'intelligible signals' several astronomers had observed coming from the Martian surface in the late 1880s and early 1890s, which some speculated had been produced by sun-reflecting communication technologies similar to the heliograph. While Percival Lowell (1855–1916) was perhaps the most enthusiastic believer that 'Martian folk are possessed of inventions of which we have not dreamed',[101] Wells responded directly to the French astronomer Stéphane Javelle's (1864–1917) observations in 1894 of a 'luminous projection on the southern edge' of Mars. In *The Saturday Review* in April 1896, Wells queried whether sentient alien life, if it even existed, would use forms of communication perceptible to humanity:

Are these senses of ours the only imaginable probes into the nature of matter? ... On either side the visible spectrum into which light is broken by a prism there stretch active rays, invisible to us. Eyes in structure very little different to ours might see, and yet be blind to what we see.[102]

Wells's interest in invisible rays would find its way into *The War of the Worlds*, with the two opposing species utilizing the electromagnetic spectrum in very different ways.

As Will Tattersdill points out, the scientist and inventor Francis Galton (1822–1911) not only hypothesized that the Earth could send out flashes to Mars as a form of interplanetary communication, but also while travelling in Africa in his youth Galton had invented a handheld sunlight-signalling device he called a 'heliostat'.[103] In using reflected sunlight as a messaging medium, Galton's heliostat was similar to the much larger 'heliograph'. This instrument was the British army's most important long-distance communications system, contributing to the maintenance and control of the colonies, and Aaron Worth has explored its presence within *The War of the Worlds*, and particularly its role in the novel's anti-imperialist critique.[104] As with the naval signal lamp, heliographs used Morse code, with short or long exposures of reflected rays of light equating to the dot and dash. There were several types of heliograph available, but the 'Mance', invented in around 1869 by Sir Henry Christopher Mance (1840–1926) of the Government Persian Gulf Telegraph Department, was the most common within the British army. It consisted of a small 4 to 5 inch wide mirror supported on a tripod stand, and could be carried and operated by one man (hence the narrator of *The War of the Worlds* comparing it to a theodolite (p. 93)). Although the use of the heliograph was of course dependent upon clear days and bright sunshine, it was celebrated for the great distances over which it could transmit, especially when positioned at a high altitude.[105] Using Morse code under the pressure of time (in the 'theatre' of war), the heliograph was limited in the subtleties of its expression, and lacked the intimacy of the telegraph. Its flashes could be viewed by anyone, with comic results in Rudyard Kipling's poem 'A Code of Morals', in which a soldier communicates nightly with his wife across the 'Hurrum Hills above the Afghan border', and his warning 'Don't dance or ride with General Bangs – a most immoral man' is intercepted by much of the British army stationed there, including Bangs himself.[106]

In a very different tone, *The War of the Worlds* is especially concerned with technologies of communication, part of a wider theme within the novel about observing and recording at a distance. The narrator is preoccupied with the management of information: how knowledge about the Martians is acquired, transmitted and received. The conventional networks of intelligence can collect and disseminate knowledge at speed, and yet telegraph transmissions, newspapers and even official reports are often inaccurate, from the narrator's own experience, as he makes the reader aware of the common circulation of spurious descriptions of the Martians and their technology. Amid these failures of media to transmit data accurately, or to function altogether, the heliograph is celebrated for its contribution to the British military's resistance against the Martians. The heliograph operators are far from being flying Apollonian messengers, but their dedication and courage is lauded:

Through the charred and desolated area – perhaps twenty square miles altogether – that encircled the Martian encampment on Horsell Common, ... crawled the devoted scouts with the heliographs that were presently to warn the gunners of the Martian approach. (p. 101)

The humans use the Sun against the Martians, not as a weapon, but as the medium of a communications network. Later, however, the reader would be excused for confusing the heliographs with some sort of armament: 'The Martians had been repulsed; they were not invulnerable. ... Signallers with heliographs were pushing forward upon them from all sides. Guns were in rapid transit' (pp. 109–10). How ironically Wells wants us to view the narrator's faith in these unreliable and intermittently functioning devices is unclear, but it is fitting that the narrator, a professional writer, places such emphasis upon the power of words.

As Aaron Worth has argued, the British army's heliograph seems to be 'a less monstrous analogue of, or precursor to, the Martians' formidable Heat Ray itself', both devices of conquest and empire. It has also been suggested that the tripods were inspired by the tripod-mounted Gregorian telescope the teenage Wells found in the attic. Coming across a box 'full of brass objects that clearly might screw together', he succeeded in assembling the device, paralleling the Martian cylinders that screw together to make a war machine with a 'brazen hood' (*The War of the Worlds*, p. 84).[107] Despite the number of parallels, this suggestion is not absolutely compelling, given that Wells would have encountered many pieces of tripod-mounted apparatus at the Normal School of Science.

The narrator is fascinated by, and gives vivid descriptions of, the Martians' machines, and in his willingness to observe them he often puts himself at risk. His first horrifying encounter with Martian technology is at the sand pits, where one of the cylinders has crashed. There he witnesses the devastating Heat-Ray, and his retrospective narrative manifests the accumulation of his experiences of this alien weapon. First, we are told that 'a humped shape rose out of the pit, and the ghost of a beam of light seemed to flicker out from it', then 'flashes of actual flame, a bright glare leaping from one to another, sprang from the scattered group of men ... as if some invisible jet impinged upon them' (*The War of the Worlds*, pp. 66, 67). Next, an 'unseen shaft of heat' bursts pine trees into fire. Soon 'this flaming death' is figured as an 'invisible, inevitable sword of heat', and 'an invisible yet intensely heated finger' (p. 67). Eventually, the narrator recognizes this 'humped shape' as a 'restless mirror wobbl[ing]' upon a 'thin mast' (p. 67), and it is here that the narrative accommodates his future knowledge:

> It is still a matter of wonder how the Martians are able to slay men so swiftly and so silently. Many think that in some way they are able to generate an intense heat in a chamber of practically absolute non-conductivity. This intense heat they project in a parallel beam against any object they choose by means of a polished parabolic mirror of unknown composition, much as the parabolic mirror of a lighthouse projects a beam of light. ... [I]t is certain that a beam of heat is the essence of the matter. Heat, and invisible, instead of visible, light. Whatever is combustible flashes into flame at its touch, lead runs like water, it softens iron, cracks and melts glass, and when it falls upon water, incontinently that explodes into steam. (p. 69)

While 'no one has absolutely proved these details' (p. 69), the narrator's account soon assumes they are true, making reference to 'the spinning mirror over the sand-pits' and the effect of the 'elevation of the parabolic mirror' upon the ray's transmission (pp. 69, 70). Even after the Martians are defeated, and their bodies and machines put under the scrutiny of human science, the narrator admits that the 'generator of the Heat-Rays remains a puzzle', too advanced for adequate explanation (p. 191). Nevertheless, the Heat-Ray weapon, as far as the narrator understands it (like the 'parabolic mirror of a lighthouse'), is a direct descendant of the burning-mirrors of the seventeenth and eighteenth centuries, and concave collectors of Mouchot and Ericsson, but with its own, artificial but sun-like power source. Indeed, the narrator's description of its powers is reminiscent of accounts of burning-mirrors one might find in the early *Philosophical Transactions* (see Chapter 2). On their home world, the Martians can no longer rely upon the cooling Sun, and so their ingenuity has successfully created an artificial 'sun-in-little' that dispenses with the real thing altogether – although it is seemingly not powerful enough to sustain life upon their planet, only conquer another. The thermal radiation it projects is detectable as merely a 'ghost of a beam', seeming to exist beyond the visible part of the spectrum, perhaps drawing upon Herschel's discovery of infrared as the hottest spectral region (see Chapter 3). Critics have marvelled at how Wells seems to have foreseen both the laser and thermonuclear energy (power that emulates the natural activities at the heart of the Sun), and created dark caricatures of contemporary heliograph devices. Yet perhaps Wells's creative act should be seen more as archaeology than futurology, as part of a long tradition of solar burning-mirrors? Indeed, at least one reader believed that the Martians 'extract from ordinary sunlight a "heat-ray" of extraordinary lethal power'.[108]

It is not just the cylinders, but the Martian tripods that are armed in this way. The tripods have been seen as a metonym for the immense power of Victorian technology.[109] There is certainly something Promethean about them, as they literalize the metaphors often applied to the Industrial Revolution: an artilleryman calls the Heat-Ray 'lightning' and a 'fire-beam' (p. 95), and the narrator refers to one tripod as a 'glittering Titan' (p. 90).[110] But the fire they have stolen is hardly a gift to mankind. Each war machine has 'a kind of arm [which] carried a complicated metallic case, about which green flashes scintillated, and out of the funnel of this there smote the Heat-Ray' (pp. 89–90). The case – soon called a 'camera' or 'camera-like generator' – is raised high in the air by the tripods when they prepare to project the 'ghostly' Heat-Ray (pp. 98, 136, 97). The devastating energy is released as a 'flash', bringing to mind the snapshot technology of artificial illumination as invented by Fox Talbot.[111] The similes equate photographic capture with a form of death. The Martians also bring with them the remarkable vegetation of the red weed, its rapid spread seeming to confirm its greater efficiency in absorbing solar energy than native species, and its colour suggesting it harnesses light across a different range of wavelengths.

The War of the Worlds can be read, therefore, as a 'whole solar drama' (to use Lockyer's term). It is concerned with the necessity and usage of solar and pseudo-solar energies, across wavelengths of light. Through shocking violence, the Martians represent a possible future result of Victorian technological ambitions, their weaponized pseudo-solar powers a distorted mirror image of coal-fuelled British imperialism, and the invasion suggesting the lengths to which empires might go to secure their sources of energy in the future, amid fossil fuel depletion and heat death. The novel's premise of the cooling Sun would soon be undermined, though. The 'Helmholtz-Thomson' contraction theory and the forecast of the Sun's lifespan as (relatively) short survived for forty years, but the discovery of radioactivity raised other possibilities. In particular, Marie Curie (1867–1934) and Albert Laborde (1878–1968) discovered that radium's decay dissipated an immense level of heat and produced helium, which in 1895 was shown to be heavily present in the Sun's chromosphere. The radiochemist Frederick Soddy (1877–1956) believed this discovery to have 'permanently altered' our outlook: 'We are no longer merely the inhabitants of a world itself slowly dying, for the world, as we have seen, has in itself, in the internal energy of its own material constituents, the means, if not the ability, to rejuvenate itself perennially.'[112] As we will see in Chapter 6, just over a decade later thermonuclear ideas transformed conceptions even further.

Selenium stories

By the time Wells was writing *The War of the Worlds*, a radically different means of obtaining solar power had already presented itself, although (as Chapter 5 will explore) its significance would not enter into popular currency until many years later. The element selenium, mainly found in iron and copper sulphide ores, had been discovered in 1817 by the Swedish chemist Jacob Berzelius (1779–1848), and was identified as having a high resistance. Over half a century later, in early 1873, Willoughby Smith (1828–91), chief electrician for the transatlantic telegraph network, was using 5 to 10 centimetre wide bars of crystalline selenium, hermetically sealed in glass tubes, to detect flaws in the cables. Investigating the great inconsistency in its operation, Smith discovered that the electrical resistance of the selenium bar varied according to the intensity of light shining upon it. Crucially, he noticed, first, that the action of light did not alter selenium's electrical properties permanently, with a slow return to its normal resistance once the light is withdrawn, and, second, that sunlight (even when apparently dull) had a much greater effect than bright artificial forms of illumination (from coal gas and other sources).[113] The Sun's artificial rivals could not compete on this occasion. Smith's accounts drew the interest of several researchers, including William Grylls Adams (1836–1915) and Richard Evans Day, whose subsequent investigations included Smith's own samples. They concluded that selenium was a 'photoelectric' substance that develops 'a kind of electromotive force' when 'under the action of light', with the change in electrical resistance directly proportional to the square root of the light's illuminating power.[114] Their work excited and intrigued the scientific community. Adams's friend and predecessor as Professor of Natural Philosophy at King's College London, James Clerk Maxwell

(1831–79), witnessed the 'conductivity of Selenium as affected by light' in April 1874, finding it 'most sudden. Effect of a copper heater insensible. That of the sun great.'[115] Maxwell, whose own work a decade earlier had redefined light, recognized that something revolutionary could be here too.

Initially, selenium's properties suggested its usefulness as a photometer, to measure the intensity of light. It was also explored, by Alexander Graham Bell (1847–1922), as the basis for the 'photophone' and, by George M. Minchin (1845–1914), for the *telephotograph* that could produce a 'photographic image of an object at a distance.'[116] Given that the cells were only 1 per cent efficient, selenium's potential for power generation took some time to be taken seriously, but by the 1890s, the 'direct transformation of the radiant energy of the sun into electrical energy' was considered 'one of the dreams of the modern scientist.'[117] Perhaps this was the means of Britain's escape from dependence upon coal which Jevons had longed for? Certainly, the potential for selenium cells to tap into a seemingly endless supply of energy was, according to the electrical engineer Rollo Appleyard (1867–1943), exciting some to behold 'the blessed vision of the Sun, no longer pouring his energies unrequited into space, but, by means of photo-electric cells and thermo-piles, these powers gathered into electric storehouses to the total extinction of steam engines, and the utter repression of smoke.'[118] This was not only a quasi-divine, sublime fantasy of energy abundance, but also a technological survival of the fittest, which coal was destined to lose. As the twentieth century dawned, science fiction began to explore the possible futures of such a transition, no matter how far away it seemed.

Bright futures:
Solar science fiction takes off

Solar fictions and frauds

The fight goes on between Dame Nature and the scientists. Whether we shall ever have an efficient solar boiler and engine is a problem worth thinking about …, as we possess no greater source of natural energy, to be had without taxation or special leases from some money-grabbing coal, oil or other baron, than that of the sun. Some day we may be able to derive all necessary light and power, for our homes at least, by means of a solar-electric plant located on the roof, and who shall say that we must be taxed for utilizing such energy?[1]

Hugo Gernsback's 'The Utilization of the Sun's Energy' (1916), written for his own magazine *The Electrical Experimenter*, raises issues that pervaded discussions of solar power throughout the twentieth century, and have continued into the twenty-first. The near monopoly held by Standard Oil in the United States had been broken in 1911, but the energy companies retained their profiteering reputation. The disruptive Pennsylvania coal strikes of 1902, meanwhile, had demonstrated the fragility of supplies.[2] Transitioning to levy-free solar power after coal, gas and oil seems not only economical and libertarian but also a logical progression, since 'what are they but transformed sunbeams?' (p. 605). Framing a speculative technological narrative also as an economic one, Gernsback sees the adoption of solar power as involving a move towards the individual producer/consumer. Imagining absolute energy justice, he assumes that solar cannot be monopolized: that the technology to harness it, like the Sun itself, will be available to all. For Gernsback (1884–1967), writing for an audience of scientists and engineers, the means of achieving this lies solely in their hands, not within a network of different agents. The Promethean, anthropocentric narrative of the individual (generally male) scientist seeking to conquer 'Dame Nature' frames much of how twentieth-century science fiction tried to understand solar energy and its technological capture.

Gernsback is a key figure in the development of science fiction, founding a whole roster of magazines (particularly *Amazing Stories*, 1926, and *Science Wonder Stories*, 1929) dedicated to publishing this genre, which he called 'scientifiction'. Moreover, many of his ostensibly scientific and technical titles also featured science fiction stories from time to time, encouraging a culture of speculative invention and futurology that hovered between fact and fiction, including in the area of solar energy. This chapter will focus on science fiction in particular: a genre that, according to Graeme Macdonald, has been 'the most reflexively and *consciously* aware of energy as literary *and* material necessity, politico-environmental issue and techno-social system', with its golden age bound up with the 'expanded and accelerated twentieth century world that oil delivered'.[3] Yet, as we saw in the previous chapter, the discourse around fossil fuels that acknowledged its ultimate origins also fomented projections of more direct solar energy capture.[4] In the solar narratives established by early-twentieth-century SF, the harnessing of this power via sublime feats of architecture and engineering functions as a synecdoche for technologically driven human futures, characterized by sustainability and/or environmental catastrophe. Solar technologies were important plot devices in the often clunky and reactionary, yet frequently visionary and technically astute, pulp SF stories. But in many such narratives, solar power scenarios operate as kinds of thought experiment that consider the possible ethical, operational, economic and social dilemmas associated with renewable and other forms of energy. Gernsback (and others) therefore created a print culture in which diverse solar energy imaginaries could flourish, encouraging not only interest and investment but also fraudulence.

Gernsback had not always sought to be a publisher. In 1904, he emigrated from Luxembourg to America, hoping to manufacture a new form of battery he had invented. When this project failed to fly, he established magazines about gadgets instead, starting with *Modern Electrics* in 1908 and succeeded by *The Electrical Engineer* and *The Electrical Experimenter*. Gernsback's publications captured a brand new phenomenon and audience, since between the 1890s and 1920s three quarters of the United States became electrified, radically transforming the country's infrastructure, landscape and way of life.[5] For Gernsback, the future seemed bright. Modernity and progress were manifested in the illuminated urban environment, a place of electrical sublimity perceived to be emulating the Sun. On a visit to New York in 1910, for instance, Ezra Pound saw 'squares after squares of flame, ... for we have pulled down the stars to our will'.[6]

For *Modern Electrics* in 1911–12, Gernsback wrote the twelve-part serial *Ralph 124C 41+*. Adopting the common motif of the foreign visitor to a land of the future, much of the story consists of the eponymous American hero showing a Swiss woman (and love interest), Alice 212B 423, around the New York of the year 2660, focusing particularly on its advanced electrical technologies. Gernsback probably saw the serialized story as a way to increase interest in the title (although many years later he confessed to have forgotten 'just *what* prompted' him to compose it),[7] and nearly every month's cover featured an illustration derived from one of its episodes. Republished as a novel in 1925, Gernsback's story has received a critical drubbing over the years, but we should perhaps view it less as a novel, and more as a kind of speculative technography (a textual world's fair, if you will) since credited with numerous successful predictions of technology.[8]

What keeps this techno-utopian, post-oil New York running is an immense solar energy station of 'twelve monstrous Meteoro-Towers, each 1,500 feet high, [forming] a hexagon inside of which were the immense *Helio-Dynamophores*, or Sun-power-generators'.[9] (This is Gernsback's coinage, literally meaning 'sun-power-bringer'.) As if this prospect were not technologically sublime enough, the narrator describes how 'the entire expanse, twenty kilometre square, was covered with glass. Underneath the heavy plate glass squares were the photo-electric elements which transformed the solar heat *direct* into electric energy' (p. 99). Each square metre contained 400 photoelectric elements, with 1,600 units placed in 'large movable metal cases' and 'mounted on a kind of large tripod in such a manner that each case from sunrise to sunset presented its glass plate directly to the sun' (p. 99). Essentially, these are heliostat-like panels which track the Sun, and are placed in such a way that 'shadows from one row could not fall on the row behind it' to ensure optimum functionality and efficiency, with each generating 120kW (p. 100). Alice remarks that they have 'similar plants across the water but [not] of such magnitude. It is really colossal' (p. 101). The scale of this development is further proof of what we gradually realize is the story's principal maxim: that 'nothing is impossible in America' (p. 118). The power station's 'monstrous[ness]' is wholly positive, enabling American materialism to thrive into the future.

To explain how such a technological marvel is possible, Ralph tells Alice a history of solar power:

> In 1909 Cove of Massachusetts invented a thermo-electric sun-power-generator which could deliver ten volts and six amperes … in a space of twelve square feet. … but it was not until the year 2469 that the Italian 63A 1243 invented the photo-electric cell, which revolutionized the entire electrical industry. This Italian discovered that by derivatives of the Radium-M class, in conjunction with Tellarium and Arcturium, a photo-electric element could be produced which was … able to transform heat *direct* into electrical energy, without losses. (p. 101)

Gernsback includes not only spurious chemical elements but also (for the first time in the story)[10] a real-life example the readers of *Modern Electrics* may well have encountered. A 'solar electric generator' had indeed caused a splash in the New York press in 1909 – the most detailed report of which came from the nature writer and novelist Winthrop Packard (1862–1943), for *The Technical World* magazine. Packard begins by acknowledging that using the Sun's energy for heat and power is 'a dream that would seem to be as utopian as any magic feat of the genie of Arabian tales', and yet had now been realized. Appearing 'before the startled scientific world' is the 'invention of a Massachusetts man, George H. Cove, which proceeds along entirely new lines and lays a simple but cunning and effective trap for the electrical energy which the sun generates'.[11] The article continues in this hyperbolic vein, and reads more like a promotional feature than an investigative report, suggesting that Packard might have fallen into a trap himself. A lengthy but opaque description of the device is given: 'The primary cell … is a three inch long alloy of several common metals, on one end of which the sun shines in a glass-enclosed space, the other end being in the shadow, in cool free air. This rod is part of a circuit wired in the ordinary way to any good storage battery.' Apparently, the temperature difference between the cell's two ends generates electricity, although this 'action' is 'not wholly understood by the inventor' (p. 357), despite Packard remarking later that the device was the 'result of many years of careful investigation and patient labor' rather than 'hap-hazard' discovery (p. 360). Like Gernsback's narrator, Packard makes sure to impress upon the reader the importance of scale, the device consisting of 'a frame very like a sash with sixteen panes, each pane enclosing the sun-ward end of sixty-one plugs, a total of nine hundred and seventy-six' (p. 357). Unlike the power station of 2660, however, Cove's device does not require heliostat tracking, as the movement of the Sun away from a perpendicular position does not cause heat loss (p. 357). Furthermore, Cove's machine will cost little more than a hundred dollars, 'is as indestructible as a kitchen range', and with two days' sun 'will store sufficient electrical energy to light an ordinary house for a week' (p. 358).

This all sounded too good to be true, and, sure enough, it was. Around that time, Cove filed a patent and was trying to raise capital from New York investors, and in October 1909 was apparently kidnapped and offered a fortune if he ceased promoting his device. This story seemed to fit with Packard's prediction that commercial power operators would be concerned by Cove's invention, since one cannot 'monopolize the direct rays of the sun' (p. 359). However, one of the alleged kidnappers claimed that Cove himself had arranged the incident as a publicity stunt.[12] Then, in August 1911, Cove and his share-selling agent were arrested for alleged fraud. Not only had Cove made false claims about the number of customer orders but also when one of his rooftop demonstration machines was examined, its storage battery was found to be powered by the mains.[13] The *Ralph* passage celebrating Cove appeared less than a month later in *Modern Electrics*, with the 'helio-dynamophore plant' illustrated on the front cover: even if Gernsback had heard about Cove's arrest, it was probably too late to amend the publication.[14]

Cove's initial success in defrauding his investors and the press was perhaps partly due to the prominence at that time of another – but in this case, genuine – solar entrepreneur, Frank Shuman (1862–1918). Shuman's designs were not photoelectric, but solar thermal/steam based. In outward appearance, however, Shuman's glass-boxes were broadly reminiscent of Cove's, perhaps explaining why Cove was able to exploit the permeable boundary between technological innovation and fraudulent fantasy. Shuman's initial hot box, inspired by John Herschel's, was 1 foot square in size and contained blackened tubes filled with ether (which has a lower boiling point than water), and once heated the ether vapour would power a small engine. A later iteration covered over 1,000 square feet and operated an engine of 3.5 horsepower.[15] Unlike Cove, Shuman had a proven track record – hundreds of patents, many involving glass manufacture – to support any grandiose claims, and by at least 1908 knowledge of his work had spread widely in popular journals.[16] His financial backing came from British investors, and in 1912 the Sun Power Company started to build an immense power plant in Egypt, just south of Cairo, for which Shuman had tested a

Figure 14 Frank Shuman's 'sun-power pumping plant', in 'A New Power for the Prosperity of Egypt', *The Sphere*, 30 August 1913, p. 243.

prototype near his home in Pennsylvania. Shuman's design incorporated mirrors to intensify the heat, and the British physicist Charles Vernon Boys (1855–1944), whom the investors brought in as an adviser, suggested using parabolic, trough-shaped reflectors reminiscent of Ericsson's concentrators from several decades before. The plant in Egypt comprised five large collectors, each over 200 feet long and 13 feet wide; the parabolic reflectors could track the sun, with gears powered by steam (see Figure 14).[17] After some false starts (the zinc boilers melted and were replaced with steel ones), the plant in Egypt was properly tested in July 1913, producing more than 55 horsepower, and exciting an audience which included Lord Kitchener.[18]

Crucially, Shuman's solar campaign was also carried out in print. Several pieces for the popular periodical *Scientific American* explained the rationale for the power plant and turning to the Sun for energy more generally, and gave detailed descriptions of his devices. Significantly, though, the majority of these were in the form of letters sent to correct what he saw as misrepresentations. Writing in February 1914, he was particularly concerned about the language of futurity that seemed to dominate discussions: 'Sun power is now a fact, and is no longer in the "beautiful possibility" stage. It can compete profitably with coal in the true tropics *now*.' But he also argued that the technology should be given time to develop: 'The steam boiler has had a hundred and fifty years' to reach '75 per cent at best', in seven years 'our method ... has already reached an efficiency of 57 per cent'.[19] A few months later, *Scientific*

American printed another negative account, based on an assessment made by Alfred Seabold Eli Ackermann (1867–1951) and read to the Society of Engineers in London. Ackermann's report was presented quite differently in *The Electrical Experimenter*, but in *Scientific American* the scalability and long-term viability of Shuman's project was questioned for a number of reasons.[20] Shuman's response not only refuted many of the claims about the inefficiency, poor construction, land occupation, maintenance and running costs of his project, but also asserted the wider justification of utilizing solar energy, in words anticipating Gernsback's: 'Sun power is most logical. We know the sun is the central power-house of our universe, and we should, therefore, attack most seriously the problem of connecting ourselves with the sun *direct*.' For Shuman, the significance for humanity could not be understated: we adopt solar power 'or else we shall have to revert to barbarism, because eventually all coal and oil will be used up'.[21] Shuman was successful in attracting interest from both the British and German governments, but the First World War disrupted his plans, and he died before the war ended. For a time, Shuman's technological ideas died with him, especially since the next few years witnessed the discovery of many oil and gas reserves around the world, but solar thermal collectors resembling his designs are a viable form of energy production in the twenty-first century.[22]

Perhaps more importantly than Shuman's technological innovation, though, he and Gernsback (inspired by the fake invention of Cove) contributed to a growing discourse of alternative energy futures based in a culture of speculative invention and fiction, with the 'construction of a commercial solar engine [being] one of the most fascinating problems that ever engaged the attention of inventors'.[23] Harnessing the Sun was reframed as an objective of industrialized modernity rather than the fantastical energy system of a mythologized past or intergalactic future, and this began to play out in fiction.[24]

Cold light

In contrast to his most famous work *Starship Troopers* (1959), the solar energy stories of Robert Heinlein (1907–88) were located firmly upon the Earth. These more domestic settings provided the narrative space to subvert, in a modest way, the gender, moral and narratorial conventions that tended to characterize mid-twentieth-century science fiction, allowing Heinlein to develop what he came to call 'speculative fiction'.[25] Heinlein's first solar power story, 'Let There Be Light' (1940), published under the pseudonym Lyle Monroe, begins as a comedy of mistaken identity, challenging our preconceptions of the image of the scientist. Neither of the main characters, the physicist Archibald Douglas and biologist/entomologist Mary Lou Martin, can quite believe the identity of the other. To Martin, Douglas looks like a 'gangster'; to Douglas, Martin looks 'Pretty fancy!'. Appropriately, the society-changing technology they produce together belies its appearance: a nondescript 'grey screen, about the size and shape of the top of card table' and made from 'common, ordinary clay', this imagined materiality about as far as possible from the technological sublime.[26]

Martin and Douglas are brought together via their interest in 'cold light' (luminescence without heat), and this cross-disciplinary partnership works to create a cheaper form of artificial lighting, by emulating the ultra-efficient, natural luminescence of the firefly, whose radiations are within only the visible region (p. 37). Despite the many examples of luminescence in nature, the firefly had long attracted special cultural fascination, scientific curiosity and had inspired attempts at technological duplication. The terms 'luciferin' and 'luciferase' (used to describe the active compounds within the firefly) date back to at least 1886, and 'luminescence' to 1888. The firefly's energy efficiency had attracted the attention of scientists such as Oliver Lodge (1851–1940), who apparently remarked, 'If the secret of the firefly were known, a boy turning a crank could furnish sufficient energy to light an entire electric circuit.'[27] Martin's interest in fireflies puts her in the illustrious company of not only Lodge but also H. G. Wells's Holsten in *The World Set Free* (1914), who as a child 'kept them in cages [and] began to experiment … upon their light'. For Holsten, however, the simultaneous 'chance present' of a 'spinthariscope, on which radium particles impinge … and make it luminous, induced him to associate the two sets of phenomena', while in Heinlein's narrative the firefly is treated as a kind of bio-solar energy source.[28] Moreover, as a kind of corrective to the atomic *The World Set Free* (and to other 'techno-utopian' narratives surveyed by Imre Szeman and Gerry Canavan), 'Let There Be Light' suggests that new forms of energy technology do not supersede their predecessors overnight, but are involved in complex, socio-economic transitional processes.[29]

While the idea of 'cold light' was relatively commonplace, it seems more than coincidental that the premise for Martin and Douglas's invention is reminiscent of Gernsback's editorials on this very topic. In more than one article, Gernsback lamented the 'criminal waste that daily occurs when we convert coal into electric light', and saw the 'great problem of the century' as the discovery of 'heatless light' that would emulate the abilities of the firefly and lantern fish. He therefore proffered that 'perhaps the final solution will be found in some device which operates by electronic bombardment of some screen of substance, which … will become intensely luminous'.[30]

'Let There Be Light' is almost unprecedented for a pulp SF short story in the amount of technical detail it contains, and is certainly pioneering in how much it derives from a female character. Indeed, Heinlein's typescript was originally rejected by the editor John W. Campbell, Jr. (1910–71) for containing 'too much girl'.[31] Martin's lesser knowledge of physics appears to be advantageous to the project, as the initial idea stems wholly from her. She relies upon analogy – wanting to 'tune the wavelength' of 'radiant energy to the visible band of frequency' like one would on a radio – to contemplate how one could potentially filter out the wastage of energy into infrared and ultra violet bands. Once the idea is established, it is over to Douglas to 'find a crystal that can be cut to vibrate' at the right frequency (p. 38). The screen they produce (after six months of painstaking, but romance-cultivating work), has an accidental but advantageous double function: it can not only convert electricity into light but also perform the reverse. They discover this after a night of intense work:

> As the first light of dawn turned their faces pale and sickly, they were rigging two cold light screens face to face. Archie adjusted them until they were an inch apart.
>
> 'There now – practically all the light from the first screen should strike the second. Turn the power on the first screen, Sex Appeal.'
>
> She threw the switch. The first screen glowed with light, and shed its radiance on the second. … He fastened a voltmeter across the terminals. … The needle sprang over to two volts. (p. 40)

Heinlein's vitalized discourse of light emphasizes the erotic parallel between the panels and their creators, but the further allegorical potential is underplayed. Like the first screen's irradiation of the second, Martin supplied the inspirational light (of knowledge) to Douglas; the second screen is then taken outside on its own, where it basks in the full glory of the sunlight, and its power-generating ability is observed and measured (p. 40). Similarly, Martin's key role in the panel's invention is increasingly obscured. At first, both Martin and Douglas humbly defer priority to each other (pp. 39–40), but Martin – while possessing far greater business acumen and political awareness – ultimately passes on all commercial decisions to Douglas. Functioning more and more as a docile (female) lab assistant to Douglas, Martin makes the coffee and marvels at her partner's intellectual brilliance – asking 'Archie, does your head ever ache?' (p. 41) – despite her own acute intelligence and crucial interventions. When the discovery is announced to the world, the news headlines read 'Genius Grants Gratis Power to Public', erasing Martin's original contribution entirely (p. 45). Heinlein might have conceded that diminishing Martin's role would help him to get the story published (given the early comments of Campbell). However, this element also conforms to Promethean expectations of the solar energy narrative: of the lone, male genius unleashing new power upon the world, solely burdened by the significance of his discovery, and punished by significant, seemingly omnipotent, depersonalized forces above him (in this case, the underhand protectionism and industrial espionage of the energy corporations). Indeed, Heinlein's typescript reveals that the story had been called 'Prometheus Carries the Torch' originally.[32] While the recurring presence of the Promethean motif is perhaps not surprising, its inflection within different stories of solar technology (and across different genres) reminds us of the complex cultural fashioning of our responses to energy systems.

Fortuitously, discovering the screen's double function seems to solve the public problem they had learnt of just the day before, when a visit from Douglas's father introduced the corporate plot: that the recent 'public utilities bill' legislated by the corrupt state government (in the pockets of the 'pot-bellied racketeers') was enabling the power companies to fix energy prices artificially high (p. 39). The inspiration for this element of the narrative can be related directly to Heinlein's involvement in the 'Ending Poverty in California' (EPIC) movement. For liberals such as Heinlein, since the Great Depression free enterprise had been worryingly replaced by private monopoly and state capitalism.[33] Heinlein was also deeply concerned about the poor state of local oil reserves, at a time of overproduction for the overseas market, especially Japan (and in January 1939, the *EPIC* newspaper had even featured an article about the potential of solar power to solve some of these difficulties).[34] 'Let There Be Light' therefore offered a speculative scenario of solar technology in which science would seemingly solve both kinds of problems. It is also a fantasy of individual invention standing up to big business, offering a technology cultivated within a domestic sphere and designed to empower the individual consumer. The Prometheus story offered a perfect symbolic paradigm, although Douglas and Martin are also like Adam and Eve, in defying authority by accessing 'secret' knowledge.

Yet the agency of science and invention, and the individual scientist, is addressed directly within the story. While Douglas thinks the current economic situation is 'way out of my field', Martin sees physics as the perfect answer: 'Dope out some way to get power without buying from them' (pp. 39, 40). It is then Douglas's turn to overreach, realizing that if the 'atonic' screen can be modified to 'vibrate to any wave length', then it can operate at 'nearly one hundred percent efficiency', meaning 'Free power! Riches for everybody! … the greatest thing since the steam engine' (pp. 41, 42). Martin brings Douglas's Promethean ambitions back to Earth, warning that the corporate monopolies would stifle such a revolutionary invention: 'Only those that fit into the pattern of the powers-that-be ever see light. … Do you really think that they'd let a free lance like you upset investments of billions of dollars?' (p. 42). This conspiracy theory from 'Cassandra' – as Douglas dubs her (p. 43) – comes true: when the screens go into production at the Douglas family's factory, shadowy forces threaten these heroes of light. Here is the central question of Heinlein's speculative fiction: Even if the technological solution to our energy problems can be found, will the demands of corporate capitalism allow its implementation? The answer to their predicament is simple but radical, once again coming from Martin: 'Give away the secret. … Let anybody manufacture power screens and light screens that wants to', and the patent will make them money enough (p. 44). Universal access to this knowledge is borne out of hostile circumstances, but for Martin its potential benefits profoundly belie these origins: 'Don't forget […] what you'll be doing for the country. There'll be factories springing up right away. … You'll be the new emancipator' (p. 45). The story's title therefore evokes both a literal and metaphorical illumination and liberation, and this democratization of energy plays out the anthropocentric first principle of the EPIC movement: 'God created the natural wealth of the earth for the use of all men, not of a few.'[35]

Sent off into a future of marital bliss, for Douglas and Martin (at least) the story is one of happily ever after. In 'The Roads Must Roll' (1940), however, we see how the transformative social energy of their invention plays out. This was the next story in Heinlein's 'Future History': a multi-narrative projection of human society over hundreds of years, in which the solar power screens played an important role.[36] 'The Roads Must Roll' forecasts a new 'Age of Transportation' in the second half of the twentieth century. Oil and coal had been 'shamefully wasted', and now huge, mechanized pedestrian roads and stairways link city and country, 'powered with the Douglas-Martin Solar Reception Screens' (note here that Martin's credit is restored). System change to renewable energy has not facilitated a 'green' utopia; the mechanized roads have encouraged a sprawling urban landscape of 'factories … covered with solar power-screens of the same type that drove the road – … hotels, retail stores, theatres, apartment houses'. Power is now in the hands of the road technicians, on whom the society now depends, with worrying social consequences, as the story explores.[37] While Heinlein's early fictions of the future do not play out as utopias, they, nevertheless, foreground solar power as the inevitable energy source for a future Earth (or at least, North America) depleted of its fossil fuels. Although solar power would not feature so prominently in later works, Heinlein retained an interest in the subject, keeping a news clipping from April 1954 about a groundbreaking 'sunshine battery' using silicon cells (see Chapter 6).[38] Perhaps, this story confirmed that his 'Future History' was already starting to come true?

Pulp fictions of power planets

Elsewhere in the culture of science fiction, solar power was taking off. As 'scientifictions' began to fill the pages of the pulp magazines, their most sublime vision of technology was probably the space solar power station (SSPS), which collects solar energy outside Earth's atmosphere (and perhaps much closer to the centre of the solar system) and beams it in concentrated form (normally microwaves) to the planet below. Reflecting and in some cases surpassing the immense power of the Sun itself, in story after story such technologies hover between providing salvation or destruction. They promise futures of infinite possibilities, but must be carefully controlled and secured to prevent malfunction or sabotage. The scenario that such power could be used for ill rather than good, or that the technology, as a fallible human product, might prove temperamental or cataclysmic, provided thrilling disaster narratives for a genre growing in popularity. Harnessing the raw power of the Sun, unfiltered by any atmosphere, became a signifier of futurity, and the technological pinnacle of any civilization. Indeed, along with conceiving animal organisms evolved or artificially enhanced to photosynthesize, Olaf Stapledon's novels *Last and First Men* (1930) and *Star Maker* (1937) take the SSPS to its logical conclusion. Stapledon (1886–1950) imagines civilizations that drain stars in barren systems as reservoirs of energy, or that surround inhabited solar systems with 'gauze[s] of light traps' to capture the 'escaping solar energy for intelligent use' – the latter concept is now known as the 'Dyson sphere', after physicist Freeman Dyson (b.1923).[39] The cosmic perspective of Stapledon's novels, however, offers little detail in describing how these systems operate on a technological or social level. In contrast, many of the short stories on which this section will concentrate provide intricate explorations.

While Isaac Asimov (1920–92) claims priority for the SSPS concept, and has been undoubtedly the principal popularizer, there are several examples predating his stories.[40] The Russian novelist Konstantin Tsiolkovsky's *Beyond the Planet Earth* (1920) imagined vast greenhouse space stations; Otto Gail's *The Stone from the Moon* (1926) envisioned large mirrors that, like burning-glasses, concentrated sunlight onto the Earth (perhaps inspired by the German physicist Hermann Oberth (1894–1989)).[41] However, Murray Leinster's 'The Power Planet' (1931), published in Gernsback's *Amazing Stories*, was probably the first to involve some sort of intense energy conversion. Over the previous two years, William Fitzgerald Jenkins (1896–1975), using the pseudonym Murray Leinster, had published several stories of villainous climate engineering involving the evil scientist 'Preston', in the pulp fiction magazine *Argosy*. In 'The Man Who Put Out the Sun', for instance, Preston threatens to block solar radiation using electrical rays in the atmosphere.[42] As Rosslyn D. Haynes notes, these stories typify a motif found in many SF stories of the period: that scientists, if they put their minds to it (and ignore the ethics), can use technology in unprecedented ways that threaten humanity.[43] Leinster's next solar energy-conscious story 'The Power Planet' (1931), however, has a much more complex premise, although it is again ultimately concerned with the problems caused by scientific discovery.

Leinster apparently said that he would 'think of something impossible, and then write a story about it'.[44] 'The Power Planet' was no exception.

The eponymous space station is a 'vast man-made disk of metal' ten miles across, and in orbit only forty million miles from the Sun. The temperature difference between the sunward and shadow sides is nearly 700 degrees, and this differential is turned into an electric current by 'Williamson cells', then transmitted by twenty-foot-long 'Dugald tubes', supplying a minimum of a billion horsepower in a beam 8,000 miles across, to be picked up by terrestrial receivers across the world.[45] 'Caldwell glass' protects those on board the power planet from the radiation (p. 199). The naming of these technologies indicates not only an attempt at authenticity but also the thrall of the individual genius as the generator of scientific knowledge, a theme that becomes a crucial plot point in the story.

The story's title has at least two meanings, since war has broken out between the nations back on Earth. The problem for the station's international crew is that their great distance away prevents them from knowing precisely what is happening back home, although they become aware of a missile heading their way. We eventually learn that what has caused the war is the 'solution of the atomic-power problem' by a 'Professor Kettle' who was then kidnapped by 'the enemy' and forced to disclose his discovery (p. 209). The Power Planet is no longer needed by the new atomic nation, and so they seek to disable all other states through its destruction. Fortunately, the incandescent Dugald beam, reminiscent of a particle accelerator, is re-programmed to destroy the missile. 'The Power Planet' reminds us that imagined futures are unlikely to predict accurately the progress of energy forms. More successfully, however, the story instantiates the political and military power of energy, the dependency of the nation state upon its energy sources, and the fragility of those supplies – issues that would come into stark relief only some years later.

Leinster's stories spawned a number of imitations, two of which appeared in the same issue of *Thrilling Wonder Stories* (August 1937). The 'Space Mirror' by Edmond Hamilton (1904–77) blurs SF with crime fiction, and opens with the arrival of Rab Crane, of 'Terrestrial Secret Service', at the 300-mile-wide SSPS, where two personnel have been found dead in suspicious circumstances. Free indirect discourse captures Crane's excitement:

> It was a staggering sight. A huge concave mirror … made a terrific shaft of incredible radiant heat that was focused upon a certain spot in the icy Antarctic. There it struck heat engines capable of generating unlimited power for Earth. It was a mighty power project, the work of years, and

would give Earth complete power supremacy in the Solar System – if it were not wrecked.[46]

That this unlimited power is the summation of labour and ingenuity over decades magnifies its sublimity, but also suggests its potential for failure or sabotage. We learn that other planets – Mars, Venus, Mercury and Jupiter – support life, and are 'old enemies of Earth, jealous of its ancient wealth' (p. 44). An attempt to murder Crane is soon made by 'tiny men' trying to 'stab long needles smeared with a shiny, sticky black' oil-like substance into his leg (p. 46) (a further example of the monstrous and 'spectral presence' of oil Graeme Macdonald has observed elsewhere in the era's SF).[47] Crane discovers that these 'bronzed, stocky' homunculi are genetically modified Mercurian agents angry with Earth because of its strict immigration policy. The Mercurians are crowded into a small habitable zone on their own planet, but banned from emigrating to the 'green, beautiful Earth' (p. 47). The 'Space Mirror' is, therefore, itself a reflection of Hamilton's United States, where the Great Depression had resulted in increased deportation of Mexicans and the establishment of border blockades between states to prevent American national migrant workers moving from urban to rural areas.[48] In retaliation, the Mercurians plan to turn the space mirror into a weapon, directing the beam onto Earth cities. Instead – recalling the resolution of Leinster's 'The Power Planet' – Crane manages to concentrate the beam onto the approaching Mercurian fleet (p. 51).

In the same issue appeared S. K. Bernfeld's 'The Solar Menace', in which a scientist–villain (in the vein of Leinster's 'Preston') called Melas Radok seeks to 'cauterize the festering mass' of humanity by scorching the Earth. Radok builds 'huge solar condensers' at the magnetic south pole, and with 'his knowledge of [magnetism's] relations with light energy' sprays a 'new magnetism … into the air to … draw the light as it was poured forth in tremendous quantities from the sun'. From his Antarctic cavern, Radok listens with delight to radio reports that weather bureaus are 'perturbed at the mysteriously daily rising temperatures'. However, the 'spell of terrific heat' ends with an 'ample supply' of rain, resulting in the greatest crop yield in history. At the moment of realizing his work 'had been benefiting mankind instead of punishing them', Radok is crushed to death by the 'imponderable mass of ice' melting above.[49]

The pulp stories of solar power, therefore, reflect concerns about energy, nationhood, the influence of the individual genius and the misuse of technology through scenarios of anthropogenic climate change. Despite the naïve and often reactionary stories which adopted the SSPS in the 1930s, they created a dystopian paradigm of solar energy which continued throughout the twentieth century. To harness the Sun's power is the ultimate ambition of not only the scientist but also the tyrant, and as the century progresses, its science fictions are more likely to weaponize the SSPS, as we will see in the following chapter. Even *Nineteenth Eighty-Four* (1949) uses the concept, when Winston Smith reads Emmanuel Goldstein's *The Theory and Practice of Oligarchical Collectivism*: Oceania's weapon development includes geological manipulation and 'focusing the sun's rays through lenses suspended thousands of kilometres away in space'.[50] While the SSPS is a metaphorical resource that defamiliarizes anxieties about other forms of energy (whose adoption seemed more imminent than solar's) and access to natural resources and spaces, it also captures something primal and sublime that cultivates less rational meanings.

The technologies within the early SF stories bear little relation to the photovoltaic cells that were the main area of solar energy research during this period. As we saw in Chapter 4, selenium's photoelectric properties were discovered in the 1870s. They could not be understood, however, until Albert Einstein (1879–1955) published his theory of 'light quanta' ('photons'). The experiments of Philipp Lenard (1862–1947), published 1902, revealed that the frequency of the light absorbed by a substance affects the energy of the electrons released, which could not be explained by the wave theory of light. Einstein showed that light contains 'packets' of energy that vary in power according to the wavelength, and that each photon knocks one electron between energy levels within each selenium atom, with the photonic bombardment forcing an electric current to flow (thereby achieving the 'photovoltaic effect').[51] This knowledge did not immediately result in better technology, however, as selenium cells were very inefficient (generally only converting under 1 per cent of the incoming sunlight into electricity) and deteriorated rapidly. Even when Bell Laboratories discovered in the early 1950s that silicon possessed photovoltaic abilities exceeding selenium's, silicon cells at that time could only convert 4 per cent of the incoming sunlight into electricity (see Chapter 6). Yet early SF embraced a photovoltaic-powered future as a profound possibility, and tropes concerned with solar energy control and consumption had a role in shaping this emerging genre. Asimov's short stories 'Reason' (1941) and 'Runaround' (1942), for example, speculated on the technological and social possibilities such a prospect might involve, thought through ideas of energy efficiency more generally, and utilized what we could call the *narrative kinetics* of the action of solar energy (in making things happen).[52] Originally published in *Astounding Science Fiction* in April 1941, 'Reason' uses a potentially catastrophic situation involving a SSPS as the premise through which to explore the laws of robotics and interrogate notions of religion and rationality. Like its companion story 'Runaround' (set in the 'Sunside Mining Station' on Mercury), 'Reason' is normally placed in a tradition of robot narratives important to the development of cyber culture, and yet it is the idea of solar energy that drives the plot and provides crucial tropes.[53]

In 'Reason', robotics technicians Mike Donovan and Gregory Powell arrive on 'Solar Station #5', where they install and test the most advanced robot yet developed, 'QT-1' (ironically 'Cutie'), with a view to replacing all human operators (at risk from 'the heat, the hard solar radiations, and the electron storms') with this model on the many orbiting energy converters.[54] However, Cutie develops a kind of nihilism that puts not only the station but also all life on Earth at risk, because it cannot believe those 'little gleaming dots' through the glass are 'globes of energy millions of miles across! Worlds with three billion humans on them!' (p. 60). Disregarding its sensory perceptions, for Cutie these 'dots' must be optical illusions, since 'when is the evidence of our senses any match for the clear light of rigid reason?' (p. 68). Thinking through the evidence, Cutie first derives that '*makeshift*' human beings, 'depending for energy upon the inefficient oxidation of organic material', could not have been responsible for its creation, because 'I absorb electrical energy directly and utilize it with almost one hundred per cent efficiency' (p. 62). Although we are compelled to laugh at Cutie, the satire works at least one other way, exploiting energy efficiency as a conceit through which to pursue a kind of *robotophilic* satire upon humanity's imperfections. For Cutie, energy absorption is the most important criterion for judging life, and so it attributes its creation to the 'Master', the solar energy converter, who 'created humans first as the lowest type, [then] replaced them by robots, [and finally] me, to take the place of the last humans' (p. 63). Both energy and religious 'conversion' are brought to the fore as Cutie, proclaimed the 'prophet' by the less advanced robots it brainwashes (p. 65), becomes subject to a form of fanaticism, self-justified through its adherence to 'reason'. Resembling a Christian sceptic, Cutie concludes that the beams are 'put out by the Master for his own purposes', since 'some things … are not to be probed into by us' (p. 68), and the uncanny, steady stare of its eyes (which are red, glowing photoelectric cells) signifies an uncompromising intellectual position. In 'Runaround', Donovan and Powell similarly encounter robots whose 'photo-electric eyes' stare at them 'unwavering and unconcerned' (p. 48), recalling burning-glass tropes from seventeenth-century love poetry (see Chapter 1).

What Powell and Donovan at first find irritating but also amusing about Cutie's delusional state (the symptoms of which seem to subsume its obligations to protect human life) soon becomes more serious when they realize an electron storm is on its way: 'Deviations in arc of a hundredth of a millisecond ... were enough to send the beam wildly out of focus – enough to blast hundreds of square miles of Earth into incandescent ruin' (pp. 72–3). Their fears of a scorched Earth are unfounded, however, as the storm passes without incident. The fanatic Cutie had, in fact, performed its original duties, keeping 'all dials at equilibrium in accordance with the will of the Master' (pp. 73–4). Powell and Donovan realize that, despite its delusion, Cutie will operate the station perfectly, managing to control the collection and dissemination of solar energy, even if for reasons different to those expected: 'He *knows* he can keep [the energy beam] more stable than we can, since he insists he's the superior being' (p. 74).

In some ways, solar energy catastrophe is merely the backdrop for the story's religious satire and explorations of artificial intelligence. However, 'Reason' is significant in the cultural history of solar power because it not only imagines a system of energy conversion that has been adopted by speculative engineers but also functions as a kind of thought experiment which brings into focus the potential ethical and operational dilemmas of such a system. Moreover, it employs as a satiric conceit the idea of energy efficiency at a time when the efficiency of real photovoltaic devices was under scrutiny. In his lifetime (1920–92), Asimov would see efficiency improve to nearly 30 per cent, particularly because of the development of the silicon solar cell.[55] But for the SSPS concept to become a reality, Asimov recognized that substantial further research would be needed. Indeed, in his novel *The Caves of Steel* (1953), the SSPS is treated with scepticism by the main characters. This is not due to photovoltaic cell inefficiencies, but because 'the speculative fringe of science had been playing with the notion for a hundred and fifty years' and had so far been unable to project a beam 'tight enough to reach fifty million miles without dispersal to uselessness'.[56] In the short story 'The Last Question' (1956), meanwhile, Asimov again cross-fertilizes scientific with religious themes, imagining a solar-powered humanity that, millions of years apart, keeps returning to the same question of how to reverse entropy and prevent the death of the stars. It is only when human civilization is extinct, and the universe itself is near to destruction, that their super computer finds the solution and pronounces 'Let There Be Light!'[57]

The first scientific paper on the SSPS topic appeared in 1968, nearly forty years after Leinster's pioneering story. The physicist Peter E. Glaser (1923–2014) first outlined the projected decline in fossil fuels and the practical problems of nuclear energy, before citing the 'major successes' of spacecraft and satellites powered by solar cells in recent years (see Chapter 6) and the 'questionable' economic desirability of collecting solar energy on Earth. To ensure a constant supply of energy, Glaser proposed using two geostationary satellites in the same orbit, 7,900 miles apart, but out of phase with one another by 21 degrees, since each of them would pass through the Earth's shadow once a day. Glaser offers some staggering figures: to have met the power requirements of the north-eastern United States in 1966 (2.5×10^7 kW), a collector area of 8.7 square miles would be needed. A dish antenna with a diameter of 2 kilometres would transmit the microwave radiation, and although objects or 'living tissues' entering the beam might incur some damage (but not 'major destruction'), the problem of safety 'should be no more difficult than that of highway and air traffic control'. While Glaser did not underestimate the task ahead to turn such a concept into reality, he concluded that it 'appears to be less of a technological gamble than it seemed when we first announced our objective of landing a man on the moon' and 'may prove to be a logical outgrowth of achievements in space'.[58] This could not be unilateral, but would require 'international assignment of synchronous orbits'.[59] Five years later, Glaser was granted the patent.[60] Others saw the potential for space-based interventions on Earth's *climate*. Self-consciously revisiting Hermann Oberth's ideas, in 1979 Krafft A. Ehricke (1917–84) proposed aiming orbital 'soletta' space mirrors at the Earth to provide light and heat outside daylight hours in order to enhance crop growth (for food and biomass) and prevent frost damage, and to irradiate solar power stations.[61] This proposal took no account of the effect this would have on global temperatures, despite the growing awareness of anthropogenic climate change (an oversight seeming to be repeated by twenty-first-century proposals).[62] In recent science fiction, meanwhile, the soletta concept has been imagined as a way of terraforming a planet too far from its sun to naturally sustain life (typically Mars).[63]

By the 1970s, Asimov's fiction was addressing fossil-fuelled ecological catastrophe in relatively oblique, allegorical ways, imagining in *The Gods Themselves* (1972) an 'inter-universe electron pump' that draws immense power from a parallel universe but which through upsetting the gravitation/radiation balance will cause the Sun to go supernova.[64] In magazine articles and interviews, however, Asimov continued to advocate SSPS development as the potential solution to what had become known as the 'energy crisis' and as a way to democratize energy control between all nations, although he makes no specific reference to the scientific papers by Glaser and others on this very concept.[65] Writing after the Organisation of Arab Petroleum Exporting Countries (OAPEC) had imposed oil embargoes against supporters of Israel, in 1967 and 1973, Asimov saw that 'world industry cannot be allowed to depend on so fragile a base' as Middle Eastern oil.[66] More positively, in an essay on solar power written for the December 1974 in-flight magazine for United Airlines, Asimov argues that, in the intervening thirty-six years since writing 'Reason', the SSPS has 'come considerably closer to practicality' (at least in terms of spaceflight having been achieved). Crucially, though, he argues the necessity for

> scientists and engineers to overcome the practical problems …, the resolution of political leaders in backing them, the ability of people generally to understand the potentialities … and [willingly] see their tax moneys used for the purpose, and, most of all, the continued stability of the world's social, economic, and technological system.[67]

Asimov's solution was more than technological, but did not involve the kind of ecologically inflected political or economic revolution proposed by some of his contemporaries.[68] However, according to Asimov, there would be significantly advantageous political by-products of building SSPS technologies, along with producing '60 times the electricity it would on Earth's surface'.[69] Satisfying collective energy needs through immense engineering feats in space would inevitably instigate international cooperation and promote world peace, with 'sufficient pride in the undertaking to allow people to think of themselves as citizens of Earth'.[70] Asimov is more pessimistic in the Reagan era, however, when he fears that the US administration will only use space for defence and espionage rather than 'solv[ing] our energy crisis'.[71]

Influenced by Asimov (and possibly Glaser), other fictions of the 1970s revisited the SSPS concept. Harry Harrison's novel *Skyfall* (1976) shares some of

Asimov's fantasies of multilateral cooperation, imagining that the energy crisis has motivated the Americans and Soviets to work together on the 'Prometheus' project: a 'machine that would circle the Earth 22,300 miles up, would reach out silver arms and seize the sun's energy and hurtle it down to Earth'. It would be the 'answer to mankind's energy problem, ... Forever'.[72] Moreover, as the American president explains in a press conference with his Soviet counterpart, in being freed from 'dependency upon our ever decreasing store of fossil fuels, ... we shall leave forever the age of suspicion and distrust between nations and enter that of mutual peace and prosperity for all' (p. 61). However, Harrison (1925–2012) is not interested in exploring the immense diplomatic feat involved in securing the international cooperation up to this point, nor is he interested in how this energy supply will encourage equality across the globe. Instead, getting Prometheus into space is the premise of a disaster narrative, with the fragility of technology the problem (perhaps inspired by the technical hitches, including malfunctioning solar panels, which beset the Skylab space station in 1973).[73] Ominously, the third-person narrator warns that the 'failure of a single component among the thousands and thousands could jeopardize everything' and if Prometheus exploded in our atmosphere, 'it would be the largest non-atomic bomb ever made by man' (p. 19). The core body of the rocket fails to separate, and much of the story is spent with the flight crew seeking to prevent absolute disaster for the 'power-hungry world below' (p. 94). Luckily, Prometheus does not crash, but neither does it make it into orbit. Yet the novel finishes on the positive note that 'soon, another gleaming speck would soar through the thin atmosphere of Earth and into space, where it would spread its silvery net to capture ... abundant energy' (pp. 269–70). Harrison's techno-utopian prospect is therefore consistent with the West's technocratic response to the 1970s energy crisis, broadly characterized as emphasizing the potential for technology to solve all problems and preserve the conditions of the industrialized economy, rather than prompting a revision of our energy consumption practices.[74]

Global warnings

A substantially different vision can be found in Arthur Herzog's SF disaster-thriller *Heat* (1977), where the SSPS is crucial to avoid environmental catastrophe. During the late 1950s and 1960s, climate scientists had been paying increasing attention to the prospect of global warming through human activity. Notably, the US National Center for Atmospheric Research was established in 1960, and hosted the 'Causes of Climate Change' conference in 1965. By the 1970s, the realization that technological modernity was affecting climate was being communicated to the public. This dissemination occurred particularly because extreme weather events, such as the droughts in the US Midwest, Soviet Union, Africa, and India in the summer of 1972, encouraged the media to consult with scientists increasingly (although the most common scientific viewpoint at this time was that the rise in dust or aerosol pollution would probably cancel out the rise in CO_2). One particular breakthrough came from the National Academy of Sciences' 'Energy and Climate' study in July 1977, which resulted in front page headlines regarding the permanent adverse effects of coal use, and (later) the passing of the National Climate Act.[75]

Heat, which came out that same year, seems to be the first novel to have imagined how to mitigate the problems of anthropogenic climate change and, in writing it, Herzog (1927–2010) consulted with a number of experts, including members of the National Academy of Sciences and the National Center for Atmospheric Research.[76] *Heat* depicts an accelerated atmospheric cycle of global warming, and in just a few months devastating extreme weather events plunge the world into chaos. Its hero is the maverick climate scientist Lawrence Pick, who battles a sceptical scientific community until the signs (unprecedented temperatures and numbers of deadly hurricanes and tornadoes) are inescapable. To Pick, the cause is certain: 'industrial pollutants, particles, chemicals [were entering] the atmosphere in ever greater amounts', with carbon dioxide emission having increased nine times since the start of the century (p. 81). Eventually, the US government not only legislates rules against its citizens' energy 'consumption rituals' (and creates 'Energy Wardens' to enforce them) but also calls in Pick to develop a technological intervention (pp. 142, 181), with thousands of workers and billions of dollars at his disposal. Pick's answer is to create a vast orbital network of mirrors which capture the Sun's rays and focus them on central receiving stations on Earth, then retransmit the heat 'by special laser beams to an orbiting emitter' (christened the 'earth-sun') which disperses it into space (p. 178). Neither Pick nor the third-person narrator ever explains why the terrestrial receivers are needed. Most glaringly, while Herzog partially adopts the SSPS concept, there is no mention of the possibility of converting the excess solar radiation into usable energy. The narrative instead focuses on the pseudo-religious sublimity of the 'earth-sun': 'a new light [which] gleamed in the heavens' and 'meant salvation', and which arrives just in time to prevent Pick's suicide (p. 190). *Heat* shows us again the cult-like, techno-utopian thrall to industrial modernity as the (under-realized) solution to the problem it causes, and an anthropocentric faith in the individual scientist as climate hero.

In 1990, the National Academy of Sciences, together with several other US bodies, made the idea of reflecting sunlight using space mirrors one of fifty-eight policy suggestions for mitigating 'greenhouse warming', although, like Herzog, they do not mention that the solar radiation could be captured and put to use.[77] More recently, however, scientists have begun to contemplate combining sunshade climate control with solar power through conversion to microwaves or lasers.[78] This dual technology is central to the Blakean graphic novel *Dark Satanic Mills* (2013), by Marcus and Julian Sedgwick (b.1968, b.1966), which posits a dystopian future England ravaged by the consequences of climate change. Vast tracts of land lie flooded, with a submerged Trafalgar Square one of the standout visions of the book.[79] If that were not enough, the thinning of the atmosphere has left the Earth exposed to deadly cosmic rays, except in those areas protected by space mirrors above (p. 13). This artificial shell covers at least much of England (we learn nothing of the world beyond that), and not only shields the inhabitants but also provides them with power: converting the solar energy into microwaves that are channelled to receiving stations on the surface (p. 53). The influence of Asimov is present not only in the choice of energy system but also in the religious mania and 'dazzled reason' that coincides with the malfunction of this solar technology.[80] This frenzy originates, though, not in the warped logic of a positronic brain but in the Blakean 'mind-forged manacles' of man (p. 22). For this England is in the grip of a fanatical church, whose leader announces that the miracle of God on Earth can be witnessed in Northumberland (p. 148). An opposition journalist realizes, however, that this phenomenon of an ever-burning light is in fact the result of a defective space mirror (pp. 133, 130). The church's live propaganda broadcast ends with the sight of the relay station destroyed by the uncontrollable force of the rays, with the pilgrims consumed by the light, their eyes caught in full shadow as their annihilation dawns upon them (p. 158).

Barring its cover, the whole graphic novel is inked only in black and white (by John Higgins [b.1949] and Marc Olivent). The chiaroscuric effects emphasize the intensity of experiences in this extreme world: of light as the agent of sublime visual, phenomenological and religious encounter; of social chaos, deprivation and tyranny; of a society caught between uncompromising binaries of belief and unbelief. Frames often slip into the nothingness of pure white or black: visual signifiers of a disturbing but inviting void that evokes both transcendence and oblivion. Ultimately, though, the book's message is not clear-cut, and does not advocate atheism nor condemn religion wholeheartedly. The anti-establishment prophet Blake, like his Romantic namesake, gives value to a kind of religiosity, but one that rejects institutional authority, and embraces the power of the individual soul (p. 153). The tyrannical church attempts to align the light with pure, unwavering obedience; 'Blake' associates it with the liberty of the autonomous self. Blake's choice, however, is to throw himself into the atomizing light, embracing the sublime fantasy of total annihilation.

Nearly a century since its conception, therefore, the SSPS continues to fascinate, allowing authors to imagine perhaps the ultimate form of the technological sublime, eliciting hope and fear in equal measure, and provoking paradoxical feelings of religiosity. The solar technologies of SF bask in the Sun and, as reflections of its glory, they form symbols of humanity's scientific and technical progress, and the light of reason. Yet, as the narratives proceed, the fantasies of control over the Sun are shown to be illusory, as the technologies malfunction, are hijacked or suppressed by malevolent agents. Solar energy comes alive as a narrative force once its relationship with humanity breaks down, forming cautionary tales of environmental incursion. More shocking are those solar technologies *intended* for destructive purposes, as the next chapter will explore.

Dark mirrors:
Solar reflections in the nuclear age

Cloudbusting

On 20 July 1936, in the ancient Greek stadium of Olympia, the beams of the noonday sun gathered in a concave glass reflector made by Germany's Zeiss Optics. They set a staff ablaze, which was carried to the altar in front of the Altis, the sacred enclosure where a flame had allegedly burnt during the ancient games. The stadium cauldron was established at the 1928 Amsterdam games, but the torch relay from Olympia to the host city, kindled by this photothermic ritual, was introduced at the Berlin 1936 Olympics and is now a key tradition within the modern Olympiad.[1] It is thought that the Greek archaeologist Alexander Philadelpheus (1866–1955) suggested this idea to Carl Diem (1882–1962), the Berlin Games organizer, who saw the sun-kindled flame as a 'symbol of purity' which resonated with other Nazi propaganda. Seeming to regret his involvement by 1939, Philadelpheus commented that Hitler had 'bathed in the eternal Olympic light, felt the instincts of his barbarian ancestors stirring deep inside him [and now] rushes forward threatening to reduce the world to ashes'.[2]

The Olympic mirror has been re-appropriated since as a symbol of the light of civilization, athletic perfection, and world unity – we all live under the same Sun. As we have seen, the heteroglossic burning-glass's fragile tension between creation and annihilation, and civilization and barbarism, is apparent throughout its history, but its destructive potential has proved to be particularly attractive to the cultural imaginary. The mythos of the solar death ray has inspired audiences for hundreds of years, but in the twentieth century especially, solar technology became a dark mirror unto itself, as an important way of exploring the problem of ultimate power. Most strikingly, new ideas of bringing the Sun down to Earth appeared in startling fashion in the aftermath of the Second World War, when atomic power was defamiliarized as a kind of uncanny other of solar, with this figuration deployed as a way of interrogating the ethical limits of scientific discovery in the nuclear age. Therefore, although we should acknowledge David Nye's notion that each type of energy system has its own 'conceptual and experiential framework', the interactions between solar and nuclear suggest (what George Lakoff and Mark Johnson might call) the 'shared metaphorical entailments' between different forms of energy, and confirm the potency of solar power as a figurative substitute for other sources.[3] These energy allegories not only lent new understandings to nuclear but also reflected back onto solar, recalibrating its own cultural meanings. As the invention of the silicon solar cell received media attention, solar technology became part of popular culture, but it has flourished there in a form far removed from the reality of photovoltaics – as a weapon of mass destruction, for audiences craving sublime spectacle with their popcorn.

Back in 1928, D. H. Lawrence (1885–1930) (channelling the disenchantment of Keats's *Lamia*), lamented that esoteric solar explorations were no more, since '"Knowledge" has killed the sun, making it a ball of gas … . How are we to get back Apollo …?'.[4] There were, however, more troubling questions associated with the source of the Sun's power. The discovery of radioactive elements prompted physicists to consider the possibility that the Sun's energy is produced by subatomic reactions. Indeed, for Arthur Eddington (1882–1944), the 'Radiative Hypothesis' seemed to offer the 'simplest theory'. Yet, if this were true, 'it seems to bring a little nearer to fulfilment our dream of controlling this latent power for the well-being of the human race – or for its suicide'.[5] Concerns emerged about the kinds of power humanity might wield as a result of thermonuclear research, and were bound up with imagery of the Sun. One of the first was H. G. Wells's *The World Set Free* (1914), which locates the origin of humanity's technological ambition in a prehistoric and Promethean desire to 'snare' the sun, and imagines its scientist–protagonist Holsten (probably modelled on Frederick Soddy) solving the 'problem of inducing radio-activity in the heavier elements and so tapping the internal energy of atoms' in 1933.[6] In the previous chapter we saw how Wells's story had influenced Heinlein's 'Let there be light', and its impact is apparent in stories included here too.

The trope of the burning-glass would also return to prominence, as writers sought to emphasize the primal powers of nuclear. Take, for example, 'At Woodward's Gardens' (April 1936) by Robert Frost (1874–1963), in which a boy torments 'two little monkeys in a cage' at the zoo by focusing a burning-glass upon them.[7] The monkeys snatch the glass, and mock-heroically the speaker describes them engaged in empirical experiment:

> [They] instituted an investigation
> On their part, though without the needed insight.
> They bit the glass and listened for the flavor. (ll. 25–7)

At the poem's climax, the satire shifts its focus:

> Who said it mattered
> What monkeys did or didn't understand?
> They might not understand a burning-glass.
> They might not understand the sun itself.
> It's knowing what to do with things that counts. (ll. 34–8)

The poem turns ironically to challenge the superiority of humankind. The boy, 'presuming on his intellect' (l. 1), might understand what the burning-glass does, but uses this tool inappropriately in tormenting this 'lower' species. The gnomic wisdom of the final line universalizes the situation of the poem and gives it an ulteriority, becoming an allegory for the exploration of wider questions: about the need for ethics to go hand-in-hand with scientific knowledge, and regarding how humankind mistreats the environment through the misapplication of technology. The poem's moral is therefore consistent with Frost's environmental sensibility evident elsewhere (particularly his belief that 'all science is domestic science, our domestication on and our hold on the planet'), and echoes his calls for a dialogue between the sciences and humanities.[8] Manipulating the Sun's rays with a burning-glass is at once one of the simplest demonstrations of the technological control of nature and yet, as Frost exploits in this poem, this act is resonant with Promethean potential in challenging the pride of humankind and the efficacy of scientific endeavour.

While all sources of energy are in dialogue with each other (since they are all mutual 'alternatives'), the metaphorical entailments that existed between solar and nuclear technologies soon became unavoidable.[9] Previous ideas of bringing the Sun down to Earth were superseded completely when President Truman (1884–1972) announced that 'the force from which the sun draws its power has been loosed against those who brought war to the Far East', reiterating Robert Oppenheimer's (1904–67) remark (adapting the Bhagavadgita) that atomic science had produced something brighter than the 'radiance of a thousand suns'.[10] The solar representation of this pivotal moment echoed through nuclear culture in the aftermath of the Second World War, with atomic power (both fission and fusion) sometimes figured as a kind of uncanny other of solar – sometimes positively, sometimes negatively. For example, the seemingly divine speaker in Randall Jarrell's (1914–65) poem '1945: The Death of the Gods' (1948) ironically identifies the 'Bomb' as an artificial Sun discovered by humanity's great wisdom.[11] Alternatively, David Dietz (1897–1984), science editor for Scripps-Howard newspapers and correspondent for NBC News, celebrated the possibilities of fostering new energy in times of peace. He predicted that atomic research would create 'artificial suns mounted on tall steel towers' to control the world's climate: 'For the first time … man will have at his disposal energy in amounts sufficient to cope with the forces of Mother nature.'[12] Domestic and commercial bliss are there for the taking, once the minor 'detail' of 'screen[ing] out' toxic radioactivity has been addressed: the 'Atoms for Peace' programme, launched by President Eisenhower (1890–1969) in December 1953, sought to turn such aspirations into reality. Based on this potential for domestication, Dietz would make optimistic claims for science more generally, seeing mankind's 'mastery of nature and of himself ever increasing' (pp. 174–5).

Ray Bradbury's 'The Golden Apples of the Sun' (1953) joins such positive representations of atomic power by imagining a time when humanity can hold the stellar body almost literally in one's hand. This short story, named after a line in W. B. Yeats's poem 'The Song of Wandering Aengus' (1899), follows a space captain's endeavours to 'touch' the sun and 'steal part of it forever'.[13] His motivation is as follows:

> The atomic bomb is pitiful and small and our knowledge is pitiful and small, and only the sun really knows what we want to know, and only the sun has the secret. And besides, it's fun, … playing tag, hitting and running. (p. 248)

The narrative ends before the sample is taken back to Earth and, as the passage above illustrates, there seems be no anxious reflection upon the risks of storing or using this power. Moreover, although the captain is described as having 'scooped up a bit of the flesh of God' (p. 248), these connotations of blasphemy do not extend to a more explicit warning. Instead, the narrator fantasizes about this wondrous near haptic, quasi-erotic, experience: 'He … shoved his fingers into the robot Glove. … the great metal hand slid out holding the huge Copa de Oro, breathless, into the iron furnace, the bodiless body and the fleshless flesh of the sun' (p. 247). Meanwhile, the captain compares his own Promethean act with the human discovery of fire, imagining how

> a million years ago a naked man on a lonely northern trail saw lightning strike a tree. … with bare hands he plucked a limb of fire, [and ran] to his cave, where he … tossed it full on a mound of leaves and gave his people summer. … And the gift of fire was theirs. (p. 247)

Bradbury (1920–2012) was perhaps inspired by a similar prehistorical flashback in *The World Set Free* (pp. 14–15), but unlike Wells's story, the primeval parallel in 'The Golden Apples of the Sun' seems to negate any apprehension about the safety or morality of the captain's pursuit.

Many other works of the early nuclear age, however, voice deep worries about this Promethean prospect, often displaced into solar narratives. So in *The Burning Glass* (1954) by Charles Morgan (1894–1958), ideas of atomic energy and weather control coalesce within a parable of solar technology, but one in stark contrast to Dietz's prophecy, inverting the naturalizing discourse which had developed around nuclear. First performed in 1953, this hybrid of melodrama, speculative fiction, espionage thriller and, above all, morality play is centred upon Christopher Terriford, director of a weather control research centre in the south of England. Terriford has discovered how to deploy his otherwise 'completely innocuous' climate device ('Machine Six') to manipulate ionized particles in the upper atmosphere to form 'lenses interposed between the sun and the earth'. These simulated lenses can focus the Sun's rays to 'increase or diminish their local power on the earth's surface', essentially creating the ultimate weapon, a 'celestial flame-thrower'.[14] While Morgan is perhaps drawing upon stories of cloud-seeding experiments, the devastating effects of Terriford's sublime technology, the eponymous 'burning glass', instantly invite comparison with the atomic 'artificial suns' that had recently astonished the world.[15] Morgan's preface on 'Power over Nature' makes the atomic analogy explicit.[16] Invoking the myth of Pandora, Morgan reflects that 'when atomic energy was let out of the box, only a few were stricken with horror at the thing itself, as distinct from the bomb that was its incidental product. The general response was wonder' (Preface, p. xv). With the 'burning glass' exercised off-stage, the play itself is more concerned with the sublimity of scientific discovery and the power of the human mind than the technology. Terriford's discovery induces a kind of trauma, as he fails to cope with his solar knowledge.

The play's crucial exchanges pass between Terriford and the prime minister, Montagu Winthrop, exploring the irony of a scientist–hero having total power but being under state control. This was a timely motif: Oppenheimer's loss of US security clearance later in 1954 confirmed that the individual scientist's place among the machinations of government was a precarious one.[17] However, even the minor characters provide different perspectives on the central question of applying science. For Terriford's mother, 'knowledge is always Progress', while the civil servant Lord Henry Strait remonstrates with Terriford that he is 'trying to change a world that doesn't want to be born again' (pp. 15, 122). The science is the product of a family venture, not an industrial complex (Terriford having inherited the Weather Control Unit and its projects from his father), and the small-scale setting belies the global implications of their discovery.

Terriford emphasizes the paradoxes of scale within the sublime prospect of the burning-glass: 'an infinite supply of pure heat, spilling out from heaven just because my puny hand turns on the tap' (p. 23). Changing the 'proportion' of things (p. 14), a single human with knowledge of the parameters can hold the power of life and death over the whole planet. Predictably, given the 'absolute mastery' he has gained over solar energy (p. 73), Terriford, the individual scientific genius, sees his situation as remarkably Promethean (Preface, p. xxvi), but his lonely existential struggle is to some extent self-imposed. However, Terriford's decision to deny others knowledge of the burning-glass's operations is suitably justified at the play's climax, when his assistant Tony Lack, having surreptitiously memorized 'the whole setting, the whole power, all, all!', commits suicide in order to be released from this 'personal hell' (pp. 141, 142). Like Faustus (or Mephistopheles), Lack is seemingly punished for overreaching himself, motivated by no longer wanting to feel subordinate to Terriford both professionally and personally (since he desired to 'know' not only Terriford's science but also his wife).

We should read the play as not only a post-Hiroshima but also a post-Operation Hurricane text, when British atomic devices were beginning to be tested, proving to the world that this fading empire was competing, experimentally at least, with the new superpowers. Britain had collaborated with the United States and Canada in creating atomic weapons during the Second World War, but the US McMahon (Atomic Energy) Act of 1946 banned multilateral development or information sharing. The Labour government committed the country to its own programme and, in October 1952, 'Operation Hurricane' detonated a bomb in the Monte Bello islands, north-west of Australia. Less than a month later, this technology was far surpassed, when the United States demonstrated its first fusion device in the 'Ivy Mike' test.[18] Nevertheless, Britain had entered the escalating atomic arms race, and was grappling with this new responsibility, engendering an anxiety about the application of science that Morgan dramatizes but extends beyond the immediate military context.

Morgan's play also emerges amid heightened anxiety about the dissemination of national secrets.[19] For instance, the defection to the USSR in late 1950 of Bruno Pontecorvo (1913–93), an Italian scientist at the Atomic Energy Research Establishment (AERE) in Berkshire, had raised the issue that, rather than nuclear technology, the human mind was the most powerful and capacious (yet vulnerable) atomic utility. The new form of mastery in the atomic age, as the *Daily Express*'s defence correspondent Chapman Pincher (1914–2014) would point out, meant that the fear of espionage was 'the price of a civilisation in which machines are so powerful that the technical facts one man can cram into his head may imperil a nation's security'.[20] Morgan's play addresses this predicament directly. Agonizing over his own sublime knowledge, Terriford decides that the full settings of the 'flame-thrower' on Machine Six should not be in the possession of anyone else, and so divides up the upper and lower intensity settings. He gives half to his wife Mary to memorize (p. 17), with the other segment in a sealed letter posted to a friend unaware of its meaning.

Terriford is a loyal British patriot who would never defect, but he disappears at the end of Act II, kidnapped by what we assume to be a communist fifth column. A copy of the device had been given to Terriford's funders, the 'Dennistoun Foundation' in California (p. 65), raising the worry that, if this version was stolen and Terriford divulged his secrets, the whole Western world would be at risk (p. 110). With Terriford absent, it is up to Mary to remain committed to his principles, allowing the *use* of Machine Six to target two areas unpopulated by humans, as a demonstration of its power (p. 108). However, Mary keeps the *setting* unknown to not only the British but also the Americans, who fear that an attack on the Terriford house would deprive them of the burning-glass capability forever (pp. 118–19). The situation is resolved when the unknown enemy realizes that Terriford is not the sole possessor of the setting, and mercifully releases him.

The Burning Glass can be placed, therefore, within a genre of imaginative works seeking to locate the ambitions, pressures, feelings and desires of the scientist at a time when the power of science and technology had seemed to increase exponentially; when recent events, to quote Truman, 'are both proof and prophecy of what science can do'.[21] This new sense of power exposed a collective 'science' to a polyphony of voices: some seeking further application of science, and championing the scientist as the new political, social and moral guide, others questioning the belief in scientific progress altogether. In imaginative works, scientists were represented within a broad spectrum, from advocates of peace to power-obsessed megalomaniacs.[22] In the theatre, particularly, the 1947 version of Bertolt Brecht's *Galileo* had introduced a more ambivalent treatment of the scientist's motives, while Hallie Flanagan Davis's *E=mc²* (1948) and Ewan MacColl's *Uranium 235* (1952) each combined caution with an awareness of the possible benefits of atomic research.[23] Atomic scientists themselves were profoundly affected by what happened at the end of the war, acquiring a new self-consciousness about their social role, and especially concerning how they would address the erosion of confidence in scientific progress. The physicists' loudest voices at this time were not fully repentant, however. Oppenheimer asserted that the 'true responsibility of a scientist … is to the integrity and vigor of his science', and the Nobel Prize winner Percy Williams Bridgman (1882–1961) pointed out acerbically, 'If I personally had to see to it that only beneficent uses were made of my discoveries, I should have to spend my life oscillating between some kind of forecasting bureau … and lobbying in Washington.'[24]

Although Morgan's play was only one of many that addressed the social responsibility of science, Terriford's particular moral position is almost unique among actual and fictional scientists. Morgan's private correspondence reveals that he knew this would be provocative: the play 'does not offer a general solution [but] ask[s] the right question'.[25] Terriford's anxiety is very similar to Holsten's in *The World Set Free*, as Rosslyn D. Haynes points out,[26] but while

Wells's protagonist sought to suppress nuclear knowledge because he saw its military potential, Terriford allows the burning-glass to be weaponized, as a form of last deterrent against a 'totalitarian enemy' (Morgan, Preface, p. xxviii). However, Terriford will not allow the peaceful use of his device because it seems that a limit has been reached. It would be a 'blasphemy of applied science' for humanity to possess such material comforts: 'Suppose we could switch on the sun to drive all our engines for us, … suppose this huge power over Nature were really on tap in every suburb from Purley to Peru – what then?' (p. 80). According to Terriford, science should be 'seen again as … a source of wisdom, not of power; … a reading of Nature, not her slave-driver' (p. 150). In denying humanity the material benefit of new knowledge, Prime Minister Winthrop points out that Terriford is 'standing against the whole tide of modern thought' (p. 149): his refusal could be viewed as an 'evil' and 'totalitarian' act itself, as contemporaneous reviewers suggested.[27] The individuality of Terriford's knowledge and moral position might have also reduced the play's effectiveness as a comment upon atomic science, practised by researchers around the world and strongly supported in peaceful applications, including medicine.[28]

Terriford's aversion to the burning-glass stems partly from his reaction at the time of discovering the specific setting. Although this moment is not dramatized on stage, having occurred six weeks prior to the action, it is discussed retrospectively during the play. Morgan's preface, which provides some of the play's backstory, also explains how Terriford had 'hit upon' the setting accidentally, not through a calculation, and yet he denied it was 'a fluke'. Instead, it was 'Total. Certain', and 'visionary' in its 'suddenness', 'completeness', and 'transcendence of reason' (pp. xviii, xix). The sublime purity of this moment drove Terriford to find 'a firm and reasoned assurance of evil intrinsic in the Burning Glass itself', a 'Devil in his toy' (pp. xvi, xxii). The Satanic language in which Terriford frames his discovery not only evokes Faustian overreaching but also reminds the audience of Oppenheimer's conviction that 'physicists have known sin; and this is a knowledge which they cannot lose'.[29]

While Terriford, and perhaps Morgan himself, found evil within the machine, William Golding (1911–93) sought to grapple with the anxieties of the atomic age by looking for signs of diabolical malevolence closer to home. Like *The Burning Glass*, Golding's *Lord of the Flies* (1954) allegorized the atom bomb as a form of solar technology. The latent evil found within the novel's juvenile characters reveals Golding's attempt to vanquish belief in progress and the innate benevolence of our species, two ideas challenged by the bombs on Japan, the post-war revelations of the Holocaust, and the increasing Cold War threat. The symbolic function of Piggy is crucial to these intentions. The first of Golding's many rationalist characters, Piggy thinks 'life is scientific' and cannot make sense of Simon's visionary capacity.[30] Golding himself described Piggy as a 'naive optimistic scientist', and this bespectacled boy's uncompromising, short-sighted belief in human progress is tested relentlessly, to a fatal conclusion.[31] At the centre of the intricate symbolic structure around Piggy are his spectacles, which physicalize his association with civilization, reason, technology and the power of knowledge. As burning-glasses, Piggy's spectacles become the essential agent of survival: the fires they facilitate provide a heat source for cooking, warmth at night, the illusion of safety in the dark, and a smoke signal out to the world. The spectacles are both a product of, and provide a signal to, civilization and yet also the means by which humanity can connect with a seemingly inherent barbarity. The narrator makes much of describing how the glasses 'flashed when [Piggy] looked at anything' (p. 81), signifying his rationalistic gaze, and the close union between Piggy, the spectacles, the Sun and human reason pervades the novel.

Lord of the Flies is in some ways a parodic rewriting of the imperialistic adventure *The Coral Island* (1858), subverting R. M. Ballantyne's reductive, racist depiction of evil.[32] Ballantyne (1825–94) equips his boys with a telescope, from which they salvage the convex lens to use as a burning-glass,[33] but Golding provides his protagonists with different means when they are desperate to start a smoke signal. It is Jack, the choirboy and future hunter–murderer, who suggests using the spectacles as burning-glasses: Piggy the 'scientist' might own them but lacks the practical insight to consider their alternative, more violent use. Snatching the glasses, Jack hands them to Ralph, who 'moved the lenses back and forth, this way and that, till a glossy white image of the declining sun lay on a piece of rotten wood'. This microcosmic reproduction does not rest for long and flames are soon flapping high from the tinder-dry limbs (p. 53). Quickly the fire is out of control, and takes on an animistic quality, 'gnaw[ing]' at the canopy: 'A quarter of a mile square of forest was savage with smoke and flame' (p. 57). The sublimity of this conflagration is soon acknowledged, with the liberty of this power a recurring trope which seeks to obscure the issue of control: 'Ralph realized that the boys were falling still and silent, feeling the beginnings of awe at the power set free below them. The knowledge and the awe made him savage' (p. 57). Piggy's spectacles, as burning-glasses, conjure up in miniature the Enlightenment prospect of technical mastery of nature, and yet almost immediately this illusion turns to smoke, and they unleash an uncontrollable and seemingly unlimited source of raw power that threatens the whole island and awakens the boys' dormant barbarity. Moreover, there are not only environmental but also human casualties of this technological intrusion, as it slowly dawns on Piggy that the 'little 'un … with the mark on his face' is missing (p. 60). Golding's fable captures microcosmically the destructiveness of humanity when it manipulates power sources, the burning-glass analogy hinting simply but audaciously that in wielding atomic energy we are but excited children. With cruel, unconscious irony, Piggy later declares that grown-ups 'wouldn't set fire to the island' (p. 117), his faith in human progress leading him to forget that the boys are probably there because of a nuclear attack back home. *Lord of the Flies* can be located alongside *The Burning Glass*, therefore, as another nuclear work which explored the character of the scientist, at a time when the power of science and technology had seemed to increase exponentially. Through Piggy, Golding's fable seems to suggest that well-meaning but naïve scientists have unleashed terrifying technologies into the wrong hands.

The novel's inconsistency regarding the spectacles is well-known: if, as the narrator reports, Piggy suffers from myopia in the technical sense, the corrective concave lens required would divert, not focus, the Sun's rays, and so it is likely that Golding meant for Piggy to be hypermetropic.[34] Nevertheless, the symbolic function of the glasses, drawing upon the conventional imaginative value of the quotidian burning-lens, transcends any technical impossibility and challenge to realism. As burning-glasses, Piggy's spectacles become the essential agent of survival, and yet they are also the means by which the boys connect with a seemingly inherent savagery. The novel's final action is a struggle between Ralph and Jack for control of the potentially destructive, but also life-sustaining technology of the glasses (a Cold War in miniature). By this point, as the boys descend further into their primitive selves, the spectacles – with one lens already broken by Jack (pp. 89–90) – even lose their nomenclature. Recalling the Promethean motif connecting the stolen fire of heaven with the Sun's power, Jack orders his tribe to 'take fire' from Ralph and Piggy, and when the attack is successful, Ralph bemoans that they 'stole our fire' (pp. 199, 209). As the object's name is forgotten, the civilization it signifies also becomes just a memory. The importance of the burning-glass in *Lord of the Flies* is, therefore, at least threefold. It reinforces the sense that some dormant characteristic in human nature has been drawn forth by the environment, with the island as a kind of lens that focuses humanity's destructive potential. Its primal symbolism contributes to the timelessness of the fable, and as a figurative substitute for nuclear technology (a kind of artificial Sun) on the microcosm of the island, it acts as a terrifying projection of future apocalypse triggered by scientifically developed powers whose dissemination humanity cannot sufficiently control.

It has been argued that the island is a paradise that comes to be destroyed by the boys, signifying man's violation of innocent Nature.[35] Yet this reading is difficult to maintain when we consider the part played by solar energy in creating the 'burning wreckage of the island' (p. 248). On numerous occasions, the Sun is described as a quasi-divine, omniscient and omnipotent, relentlessly penetrating, oppressive thing of 'enmity' that 'gaze[s] down like an angry eye' in order to impede the boys (pp. 20, 74). While critics have made much of the moral of this fable – that the boys externalize into the devil and his agents an evil that is found within themselves – Golding's narrator is preoccupied with convincing the reader of the Sun's malevolence and material presence, maintaining a sense of equivocation. Indeed, it is the character who acknowledges this internal 'beast', Simon, upon whom 'the arrow of the sun' falls, anticipating his murder during the frenzy of the hunt (p. 164). As the novel progresses, the Sun also associates itself more and more with Ralph, including when its light 'blink[s]' down while he evades capture, prompting him to think that the flickering was in his own brain (p. 243). These episodes of uncanny solar presence go beyond mere pathetic fallacy: they infiltrate and disturb the novel's wider moral schema. The boys vainly attempt to control the Sun through technology, while the narrative fails to contain or adequately explain the agency of solar energy, reflecting the moral and intellectual uncertainties of atomic power that still haunt us.

The violent and atomic associations of solar energy are also at the forefront of Arthur C. Clarke's 'The Stroke of the Sun' (1958). Of all the SF authors of the mid-twentieth century, Clarke (1917–2008) is normally claimed to be the futurologist par excellence, famously proposing the use of solar-powered geostationary satellites for television and communications.[36] A prolific journalist and novelist, across various genres Clarke entertained the prospect of a solar-powered future, and populated his work with both 'fact' and speculation. The Sun itself was a source of fascination for Clarke: a body that constantly 'communicates' with us, sending out radio waves (discovered only in 1942), and since 1959 offering a whole new form of astronomy. For Clarke, it is a place of not only 'awe-inspiring' but also 'inexplicable' forces, as he compares nuclear fusion to a kind of 'solar transmutation'.[37]

Despite realizing that 'all the other known energy sources are millions of times weaker than sunlight', Clarke was not an advocate of renewable energy to the extent of Asimov. However, Clarke made some modest but symbolic gestures, such as powering his Sinclair C5 (that 1980s icon of commercial failure) around his neighbourhood in Colombo, Sri Lanka, using a solar cell.[38] In various essays, Clarke expounded that although the Earth receives solar energy 'equivalent to a one-kilowatt electric heater on every square yard of our planet's surface', this is 'only a minute fraction of the Sun's rays; most of the energy goes rushing past into space and is ... completely wasted'.[39] For Clarke, the solution (familiar to readers of Asimov and pulp SF) was technologically difficult but conceptually simple, and provided another reason for space exploration: 'Set up light-traps very close to the Sun, and beam the resultant energy to the points where it is required'. However, Clarke was writing at a time when the Bell Labs' silicon cells had only just been developed (see next section), with low efficiency (under 10 per cent) and high production costs (around $100,000 per horsepower), meaning that a '100 hp automobile would require about 1,000 square yards of collecting surface – even on a bright, sunny day'.[40] For Clarke, the future – including the future of energy – always lay in the stars. Indeed, one of his most famous stories of solar energy was 'The Wind from the Sun' (1963). Later republished as 'Sunjammer', this story imagined a race between sailed spacecraft powered by the solar wind, the 'general efflux' of charged particles released by a star's corona, identified by Eugene Parker in 1958, and confirmed by spacecraft in the early 1960s.[41] The story itself is focused more on the heroism of the pilot John Merton (perhaps inspired by Clarke's meeting with Yuri Gagarin in 1961) than on the sail technology, but the concept has since inspired space engineers, resulting in several prototype vehicles, including NASA's cancelled 'Sunjammer'.[42]

For all of Clarke's celestial ambitions, his first fictional foray into solar power was located firmly on the ground, in both its setting and comic tone. Clarke had read voraciously the pulp magazines of Gernsback in the 1930s, and knew the work of Asimov and Stapledon,[43] but wanted to adapt the dreaded solar death ray or sublime solar power station for his own, half-ironic photonic fictions. In 'The Stroke of the Sun' (1958), Clarke instead imagines an ingenious way to register one's protest at a football referee. The short story's narrator, it emerges, is an arms dealer to the fictional South American country of 'Perivia'. Invited to the annual game between the national football team and its neighbour 'Panagura', he notes that it was well known that the referee at their previous clash had been bribed by the Panagurans. The same official, dressed in a 'bulletproof vest' as a precaution, would be adjudicating this time too, and the spectators, many of them from the military, had been searched for 'concealed firearms' as they entered the stadium.[44] However, 'one of the country's leading industrialists' supplied 50,000 free copies of the 'Special Victory Souvenir Issue' programme, 'bound in metal foil that gleamed like silver' (pp. 72, 73). At a bugle call, the supporters are called to action, and the narrator shockingly describes how 'the vast acreage of faces opposite me vanished in a blinding sea of fire' (p. 76). The narrator's initial reaction foregrounds the nuclear anxiety that annihilation could be always just a second away, but is undercut by a blackly humorous demise:

> I thought of atomic bombs and braced myself uselessly for the blast. But there was no concussion – only that flickering veil of flame that … vanished as swiftly as it had come … . Where the referee had been standing, there was a small, smoldering heap, from which a thin column of smoke curled up. (p. 76)

Soon, 'in a flash of incredulous comprehension' (appropriately enough) the narrator understands (p. 76):

> SELDOM do we realize just how much energy there is in sunlight. I've since looked it up, and the experts say that more than a horsepower hits every square yard of the Earth. Those fifty thousand well-trained fans with their tin-foil reflectors had intercepted most of the heat falling on one side of that enormous stadium – and aimed it all in one direction. Even allowing for the programs that weren't tilted accurately, the late ref must have absorbed the heat of about a thousand electric fires. (pp. 76–7)

The narrator's explanation is characteristic of Clarke's 'tall tales' of science fiction composed at this time (many of them collected in *Tales from the White Hart*, 1957), imparting a scientific fact while indulging in the schoolboy thrill of taking it to its *reductio ad extremum*.[45] This story apparently highlights how something so 'natural' can be turned into a weapon so easily, using simple, seemingly innocent materials, and working to such precision. As Adam Roberts notes, because the reader would know that Clarke is such a meticulous mathematician and engineer, they would be fairly certain that human combustion might be possible by these means.[46]

The man's combustion highlights the power of scale, and its ability to go beyond the imagination: just one metal foil reflector achieves very little, but is apparently devastating when in combination. Herein lies the story's political message, with the smuggling of this weapon into the stadium forming a microcosmic mirror of the narrator's own profession. Yet the narrator of this structurally ironic story seems unaware or unable to admit that his own dealings also have destructive effects, compounded by the condescending tone in which he discusses the hostilities (and political tyranny) between and within the two nations. In the kind of source his narrator might have 'looked up' the topic of energy in sunlight (which, as a metafictional joke, would include Clarke's own essays), one can find ironic treatment of their technological ambitions:

> Most new-fledged solar energy hunters are smitten with the idea of concentrating sunlight with mirrors. A horsepower per square yard looks very attractive as an inflow of power. Let's put up a reflector 10 yards on a side and concentrate 100 horsepower to run a steam engine! But large mirrors are costly and fragile, must be kept clean and free of dust, and must be turned to hold the sun's image still as the earth rotates.[47]

While Clarke was himself optimistic about the prospect of scientific progress, 'The Stroke from the Sun' is a self-knowingly tall tale that, in its overestimation of the effect of concentrated sunlight, serves as a cautionary story for his contemporaneous solar energy enthusiasts, anticipating the subversive and satirical responses of the more recent past explored in the next chapter.

Silicon heroes

Just at the time when Morgan and Golding were composing their anti-nuclear parables, another challenge to atomic power was emerging. Two scientists at Bell Laboratories, Calvin Fuller (1902–94) and Gerald Pearson (1905–87), were experimenting with impurities within circuits of silicon, and discovered that lamplight caused electricity to flow. The electrical engineer Daryl Chapin (1906–95) joined the team and made many modifications, producing a photovoltaic cell with double the efficiency of selenium. The Bell Solar Battery was presented to the press on 25 April 1954, with lamp-powered cells moving a model ferris wheel; a radio transmitter powered by the battery was demonstrated in Washington the next day.[48] The *New York Times* reported that 'this modern version of Apollo's chariot' could 'mark the beginning of a new era, leading to the realization of one of mankind's most cherished dreams'. The article's title ('Vast Power of the Sun Is Tapped by Battery Using Sand Ingredient') emphasized the abundance and ordinariness of the raw material (although its mining is not without risks to human health). The strips of silicon 'can deliver power from the sun at the rate of 50 watts a square yard of surface', apparently 50,000,000 times the power of the recently announced atomic battery from RCA (Radio Corporation of America), which was also marketed as harnessing the Sun – another example of the representational clash between the two forms of energy.[49] Countering the power of the atom, Bell Labs promoted their own device as not only a 'glimpse [of] exciting things for the future' but also one that used the 'same kindly rays that help the flowers and the grains and the fruits to grow'. Moreover, their work was presented as the culmination of a great body of research of ancient pedigree: 'Ever since Archimedes, men have been searching for the secret of the sun.'[50]

Bells Labs' discovery coincided with concerted efforts by US solar energy researchers to secure public and private funds, but the inexpensive price of fossil fuels and the assumed potential of atomic power scuppered them.[51] However, the likely usefulness of photovoltaics within the growing space race was then realized. When, on 30 July 1955, the *New York Times* reported President Eisenhower's announcement that America would send up a satellite, it included an illustration showing the solar cell power source. The first to use solar cells was Vanguard 1, launched in March 1958, and still functioning four years later.[52] While solar usage excelled in space where robustness and efficiency took precedence over cost (with around a thousand American and Russian spacecraft using solar cells by 1972), the more stringent demands (such as resistance to radiation damage) placed on solar cells drove up their costs universally, affecting their take-up in Earth-based technologies.[53] At the same time, however, the association of solar cells with space exploration was deployed rhetorically by those advocating domestic use, to express excitement and as a way of explaining their many advantages both in space and on Earth. Given they 'have no moving parts, consume no fuel, produce no pollution, operate at environmental temperatures, have long lifetimes, require little maintenance, and can be fashioned from silicon, the second most abundant element in the Earth's crust', to invest in further research was put forward as a no-brainer, even if the cost was then preclusive.[54]

The silicon solar cell seemed to bring a new urgency to thinking through the Sun's power, and particularly in relation to humanity's energy needs and the environment. Wallace Stevens (1879–1955) had been born in the decade selenium's photoelectric properties were recognized, and would die a year after silicon became the new solar material of choice. The composition date of one of his most famous pieces, 'The Planet on the Table', is unknown, but assumed to be 1954, written just in time for the publication of his *Collected Poems*. Its central premise is one John Tyndall would have approved of, that the Earth, its life forms, and all of their activities and products (poems included), originate in the Sun.[55] Poetry, therefore, must reflect this commonality, an anti-anthropocentric avowal that has been celebrated by ecocritics since.[56] The turn towards solar energy in the wake of the silicon cell's discovery is also apparent in 'Solar' (1964) by Philip Larkin (1922–85). Here, the Sun is in the organic image of one of the life forms it sustains, a 'Single stalkless flower' which 'give[s] for ever'. This eternal abundance is fortunate, since the speaker recognizes how 'Our needs hourly / Climb'.[57] Similarly reinforcing the importance of this 'Flower on its own, without a root or stem' is 'Sunlight' (1971) by Thom Gunn (1929–2004). Unlike Larkin's poem, however, Gunn's recognizes that the Sun's 'concentrated fires' will outlast us but are not eternal, only an 'image of persistence'.[58] Reflecting upon the 'passionless love' received by the Earth ('Sunlight', l. 26) allows both Larkin and Gunn to combine the sublime with the everyday, the poems themselves attempting to capture the Sun's power on paper.

The influence of the silicon solar cell, and especially its early adoption within space exploration, is more discernible within popular culture. Perhaps most strikingly, the solar source of Superman's powers was established in 1961, over twenty years after Joe Shuster (1914–92) and Jerry Siegel (1914–96) had created their hero. It was Otto Binder (1911–74) (who had initially written for pulp SF magazines in the 1930s) and the illustrator Al Plastino (1921–2013) who were responsible for the revelation that Kal-El's enhanced optical and mental abilities were due to the 'ultra solar rays of Earth's yellow sun'.[59] Since then, Superman has become a kind of living solar battery, recharging or boosting his powers by basking in sunlight. Kal-El, an alien from the planet Krypton, is naturalized (somewhat paradoxically) as a solar cell made flesh, and his strict moral code reflects the purity of the light he absorbs. Superman is only the first in a long line of solar-powered heroes, the second most famous being Cyclops, leader of Marvel's X-Men. The technologized explanation of Superman's abilities is implicit, while in 1968 the optic-blast eyes of the mutant Cyclops are explained as 'miniature, yet ultra-effective solar batteries – storing up light as do the plants about us ... and, in a process not unlike photosynthesis, they convert it into irresistible energy!'[60]

Even the Daleks were solar-powered in their second *Doctor Who* story, *The Dalek Invasion of Earth* (1964). Originally sustained by static electricity on their home planet Skaro, the Daleks' terrestrial conquest in the twenty-second century is made possible by parabolic solar collectors on their backs.[61] It was not until *The Enemy of the World* (1967–8), however, that human-wielded solar power occupied a central role in a *Doctor Who* plot. Set in 2018, the world is controlled by the United Zones Organisation, under the sway of the Mexican businessman Salamander, whose status and wealth has been accrued by increasing crop yield via concentrating the Sun. Although never fully explained in the story, Salamander also seems to use his 'Mark VII Sun-catcher' to disrupt the agricultural output of other regions, and to engineer volcanic eruptions which devastate rebellious populations (reminiscent of the dastardly climate engineers of the 1920s and 1930s SF stories). Indeed, Salamander compares an extinct volcano to 'a man in the hot sun, sleeping. Still, lifeless. Then, boom! He awakes, full of energy.'[62] The budgetary restraints and technical limitations of the 1960s BBC left these solar manipulations all off-screen, however, adding to the mystery of Salamander's device. In *The Seeds of Death* (1969), another *Doctor Who* story set in the twenty-first century, the Martian Ice Warriors plan to invade the Earth via the 'T-Mat' teleportation system, maintained by a solar-powered relay station on the Moon. The Doctor is famous for his/her abhorrence of weaponry, but this story is notable in the programme's history because it is the first time the Time Lord wields a gun: a parabolic solar collector he has adapted to project rather than receive energy, burning the marauding cold-blooded Martians to death.[63] Solar power returned as a major plot motif in

The Ark in Space (1975), set many thousands of years in the future, on a space station holding the human race in suspended animation while the Earth is bombarded by solar flares. The 'solar stacks', the Ark's main power source, are disabled by alien insects the Wirrn, whose pupae feed off the stacks directly: one of several characteristics which demonstrate their physical superiority over the human race in this struggle-for-existence plot.[64] *The Ark in Space* perpetuates the fear, instilled in Wells's *War of the Worlds*, that solar power is used more effectively and efficiently by the non-human, hostile species, since the uncanny Sun is ultimately unknown and alien to us. For a popular TV programme on a shoestring budget, the concept of solar power was a world-building element able to generate a sense of futurity and sublimity within the audience's imagination, without needing lavish props and sets or lengthy explanatory dialogue.

As a potent trope signifying ultimate power, ambition and tyranny, the solar death ray has also been assimilated into a long line of fantastical, nuclear-substitute weapons wielded by Bond villains. The film of *The Man with the Golden Gun* (1974), directed by Guy Hamilton (1922–2016), was made at the height of the 1970s energy crisis (see Chapter 5). Britain was particularly prone to the effects of OAPEC's embargo as it was also experiencing a coal shortage, and the major North Sea oil and gas fields discovered in the late 1960s had yet to come on line. To the original Ian Fleming (1908–64) novel and its initial script treatment by Tom Mankiewicz (1942–2010), the returning Bond screenwriter Richard Maibaum (1909–91) added the plot concerning the whereabouts of a microchip device, the solex agitator.[65] This pocket-sized instrument is the vital component within an advanced form of solar collector that is apparently 95 per cent efficient and which, it is claimed, will solve the current energy crisis. The professional hitman Scaramanga double-crosses his employer Hai Fat and takes the device for himself (making him another Promethean stealer of the Sun's power), planning to auction it to the highest bidder, who – as Bond suggests – could turn out to be oil suppliers seeking to suppress, rather than exploit the technology. At the film's denouement, on a secret island hideout, Scaramanga is the man with *two* golden guns, using the device to power turbo electric generators and a deadly laser. Much is made of Scaramanga having the Sun in his pocket, emphasizing not only the micro-sublimity but also the phallic symbolism and fetishization of handheld technology within the film as a whole.[66]

Die Another Day (2002), another of Bond's more kitsch outings, recalls Morgan's *Burning Glass* in defamiliarizing the 'rogue state' nuclear threat through solar technology. As a way of celebrating Bond's forty years on screen, the writers Neal Purvis (b.1961) and Robert Wade (b.1962) sought to capture the scale of the early films, with a weapon reminiscent of the laser satellite of *Diamonds Are Forever* (1971).[67] A North Korean colonel, Tan-Sun Moon, has developed an orbital solar mirror in order to cut through the minefields of the Korean demilitarized zone, an area haunted by the threat of nuclear war. Disguised as British philanthropist Gustav Graves, Moon presents his project publicly as a beneficent, beautiful piece of technology which will supply all parts of the world with light and heat, and tackle famine by increasing crop yields. The film's contemporary energy context is less developed than *The Man with the Golden Gun*'s, however; dispensed with in Moon's throwaway sarcasm about global warming, as he wields his heat ray. Partly due to the overblown *Die Another Day*, the solar death ray is now probably the tech-villain's most clichéd weapon of mass destruction. This is crystallized in the parodic animation *Megamind* (2010), where the eponymous dastardly genius seeks to atomize the hero Metroman, but waits frustratingly for a cloud to blow out of the way. *Star Wars: Episode VII – The Force Awakens* (2015), however, returned some of the solar death ray's terror in the form of 'Starkiller Base', a descendant of the original film's Death Star which, rather than emulating the Sun's power, drains a star's energy completely and converts it into a weapon able to destroy several planets simultaneously.[68]

Sublime or ridiculous, these popular entertainments have kept the idea of solar energy prominent in the public consciousness, normally as a figurative substitute for nuclear, reminding us of the respect humanity must give to the powers it unleashes. Destructive rather than generative, and often self-consciously exaggerated, these fictional versions of solar devices have nevertheless served to demonstrate to audiences the immense energy one can harvest from the Sun, even if the technologies themselves do not (yet) exist. More recent examples prompt the consideration of the unintentional ways humanity is enhancing the Sun's powers upon Earth artificially, through the production of greenhouse gases. This ultimate irony, of solar energy being an agent of our salvation or destruction, is at the forefront of the final chapter.

Self-renewable:
The satire and psycho-thermodynamics of solar

Hot properties

The thirty-seven-storey skyscraper at 20 Fenchurch Street, London, known as the 'Walkie Talkie' (Figure 15), is curved to maximize rental space at the top of the building, but as a motorist discovered to his horror in September 2013, under the right conditions this glass parabola can act as a burning-mirror. Reflecting the Sun's rays into a focus on Eastcheap road below, the then-unfinished building briefly became known as the 'Walkie Scorchie', melting side-panels, a wing-mirror and badge on a parked black Jaguar car (with something gloriously appropriate about solar power damaging a vehicle infamously alleged to be a gas-guzzler). The UK media were soon to seize upon this story, and thousands of online commentators vented their disbelief at the architect's disregard for 'basic' physics or, in equal measure, expressed their delight at the stuff of science fiction or Bond films coming true. Some even wondered why buildings are not designed intentionally to collect the Sun's energy in this manner. The skyscraper's developers explained that the building's ability to scorch would be a rare occurrence, happening only two or three weeks a year and for only a couple of hours each day, due to the Sun's particular elevation. They were nevertheless working on a permanent prevention.[1] It came to light that the architect Rafael Vinoly (b.1944) had previously designed the similarly vitreous Vdara Hotel & Spa on Las Vegas Strip. This building is also known to produce an occasional photothermic effect, dubbed the 'Vdara death ray', but referred to by the hotel's parent company as the wonderfully understated 'solar convergence'.[2] Interviewed about the Walkie Talkie's photothermic ability, Vinoly admitted that computer modelling suggested this could happen, but

Figure 15 20 Fenchurch Street, the 'Walkie Talkie' building. Adobe Images.

had underestimated the temperatures it could achieve. Tentatively, he claimed climate change to be responsible: 'When I first came to London years ago, it wasn't like this … Now you have all these sunny days. So you should blame this thing on global warming too, right?'[3]

The British comedian Stewart Lee (b.1968), ending one of his BBC2 shows with a routine about the disparity between London's rich and poor, mentioned the 'Walkie Scorchie' story and commented: 'as a piece of architecture, that is abysmal, but as an extremely heavy-handed satire of exactly where we're going wrong, superb'.[4] Lee did not explain what he meant, but presumably had in mind a dystopian scenario of the rich literally using their propertied power to destroy the inhabitants below. The technologically sublime skyscraper is a symbol of social and economic power,[5] a symptom of our collective vitromania (obsession with glass) and a monument to human ingenuity. But the Walkie Talkie's photothermic abilities bring to light inescapably the exploitation of nature such buildings embody, and the waste of 'good' solar energy that could have been harnessed for constructive purposes, rather than generating its death-ray phantom other. Essentially, this 'side effect' story highlights that humans fail to take fully into account how their technologies will interact with the environments in which they are placed, that they neglect the power of 'nature' at their peril, and should instead be working synergistically with the environment. It is truly a tale for our time of the Anthropocene.[6]

The Walkie Scorchie story, therefore, can be read as a kind of metaphor for the climate crisis and our sense of helplessness in the face of it. Both seem to involve corporations ignoring the available science, putting profit above environmental effects and squandering opportunities to be sustainable; governments allowing this to happen; and public exasperation at the way in which large companies and institutions conduct themselves, and over which the individual has no ability to control. As Timothy Clark has shown adeptly, the detrimental 'scale effects' of climate change negate individual agency, especially in terms of reducing material and energy consumption.[7] Much of this chapter is concerned with fictional solar energy narratives which raise questions about individual or communal agency not in relation to consumption, but instead how one (and particularly the single scientist and/or writer) enables or encourages the decarbonization of energy production. The works explored here were all published since the 'energy crisis' of the 1970s, and amid the growing prominence of the environmental and sustainable energy movements that began to emerge around the same time. An increased awareness of anthropogenic global warming or climate change exacerbated those anxieties about energy production and consumption. Although it was proven in at least the 1950s that human activity would raise global temperatures considerably, it was not until the 1980s that the increased greenhouse effect became wide public knowledge. The United Nations (UN) established the Intergovernmental Panel on Climate Change in 1988, with its first assessment report in 1990. Significant milestones since then include the 1997 UN Conference on Climate Change in Japan, at which most nations adopted the 'Kyoto Protocol', committing them to control emissions of the main anthropogenic greenhouse gases; the 2016 Paris Agreement, which seeks to limit global temperature rise this century to below 2 °C, and the 2018–19 school climate strikes led by Greta Thunberg.[8] However, most climate scientists argue that current mitigations of climate change do not go far enough, and major changes to our ways of life are necessary.[9] Moreover, the situation has not been helped by major Western leaders who, influenced by the concerns of corporations, have allegedly sought to 'get rid of all the green crap' (renewables subsidies) from energy bills, or stated publicly that the 'big push to develop alternative forms of energy … is really just an expensive way of making the tree-huggers feel good about themselves' (unless it could help pay for a border wall).[10]

The last forty years have also seen significant advances in solar photovoltaic (PV) research. These include the development of thin films (a process which involves spraying surfaces with photovoltaic substances, thereby eradicating the need for lengthy and expensive preparation of crystals), triple-junction cells, and the use of alternative materials to silicon, such as cadmium, tellurium, indium and gallium.[11] In the last decade or so, researchers have developed 'perovskite' solar cells with active layers made from the compound methyl ammonium lead iodide (MAPI). The cells have low production costs, and efficiency has increased rapidly since the initial prototypes (from around 4 per cent to over 20 per cent, between 2009 and 2016). 'Tandem' cells, which involve spraying a silicon cell with a thin layer of perovskite, are 25 per cent efficient. PV currently supplies around 2 per cent of the world's electricity, and although predictions based on cost suggest it could be the dominant provider by the 2040s, this will not become a reality without significant societal and behavioural changes.[12]

Particularly since the late 1990s, solar energy-conscious writers have been aware that the adoption of non-polluting, renewable forms of energy should be part of wider collective social transformations necessary to reduce global warming. Yet the majority of the works considered here have not been composed to shock audiences through horrifying dystopian visions of the imminent future. Instead, they draw upon ironic and satiric modes that express frustration with the lack, or at least ineffectiveness, of collective action to date, and seem designed to appeal to a more cynical age that is not suspicious of science to the point of scepticism, but aware that science and technology cannot provide an environmental panacea.

Peripheral visions

The clinical violence of the built environment reified (ray-ified?) in the 'Walkie Scorchie' sounds like something out of a story by J. G. Ballard (1930–2009), and is perhaps most reminiscent of 'The Ultimate City' (1976). This dystopian short fiction is set in a post-fossil fuel future, when conurbations have become abandoned wastelands of metal and glass.[13] The protagonist Halloway grew up in 'Garden City', 'the first scientifically advanced agrarian society', Ballard's vision chiming with contemporaneous environmentalists who thought that 'alternative' energy technologies 'can only be developed – at least on any significant scale – within the frame work of an alternative society'.[14] Its eco-community had built a 'pastoral paradise' in which 'each home was equipped with recycling and solar-energy devices' and 'powered and propelled by a technology far more sophisticated in every respect than that of the city they had abandoned'.[15] Finding it culturally stagnant and paralysing, Halloway rebels against this decentralized, rural eco-life, using a glider to escape to the deserted urban environment where, as a kind of post-industrial Robinson Crusoe, he tries to 'restart the city' using leftover fuel and still-functioning electricity generators and motor vehicles (p. 52). Some members of the eco-communities do follow him, but this new urban society soon descends into chaos, and critics have tended to concentrate on the repeated failure of modernity this enacts, and on the story's parallels with *The Tempest* and Defoe's solitary hero.[16] The title of the story, therefore, functions as a puzzle regarding which, if either, of these cities is the 'ultimate'.

The story hangs, however, on Halloway's childhood encounters with two forms of energy (one hydrocarbon, the other solar), both of them contributing to driving Halloway away from Garden City. Coupled with 'psycho-geography' (Ballard's typical interest), then, is an exploration of the psychology of energy encounters, what we might call 'psycho-thermodynamics'. The first experience is an exhilarating, visceral buzz, but with long-lasting mental consequences. Entering the workshop, Halloway finds his father running an old petrol engine, and is taken in by the 'overwhelming energy of this machine', its 'violent but controlled noise', 'juddering motion' and 'heady fumes' which 'almost knocked him off his feet' (p. 16). By contrast, Halloway's experience of solar is psychologically devastating (but, to the reader, blackly comic). His parents had burnt to death in their sauna, 'by a bitter irony killed ... by the overloaded circuitry of an advanced solar device' his 'endlessly inventive' father had designed. Rather than a symbol of hope and human ingenuity, the solar energy rig was made of 'hundreds of occluded mirrors ... fused by the intense heat of the fire, [which] towered fifty feet above [as] an all too melancholy memorial' (p. 9). The kind of parabolic mirror system Ballard perhaps has in mind is a scaled-down version of the Odeillo solar furnace in the Pyrenees. It consists of over 8,000 individual mirrors (covering 2,400 square yards) with, crucially, sixty-three electronically controlled heliostats which track the Sun in order to focus upon the main section. Designed by chemist Félix Trombe (1906–85), based on his smaller furnace from the late 1940s, it was built during the 1960s, and operational from 1970. The furnace was (and still is) used for conducting high-temperature experiments, not power generation, but its ability to reach temperatures of up to 3800°C and produce power greater than 1,000 kilo Watts (now much higher) was used in popular works on 'alternative technology' as evidence of the immense potential of solar energy.[17] The sublime appearance of the furnace, a 'silvered glass in the green valley … nine stories high', helped to convey the power at its disposal.[18] Ballard, however, fosters an alternative aesthetics within such structures.

Coupled with (and perhaps because of) this childhood trauma, in Garden City Halloway sees only 'boredom and indifference', its inhabitants 'casual[ly] winning' energy 'from the sun, the wind and the tides', as if, in a warped logic, energy is only morally acceptable if derived from extractivist labour (p. 16). There are even some members of the community who take their solar enthusiasm to its *reductio ad absurdum*: 'a fringe group of scientific fanatics – the so-called "heliophiles" – whose ambition was to return energy to the sun by firing off all the old missiles with nuclear warheads, repaying the sun for its billion-year bounty' (p. 45). Halloway fails to recognize the energy monomania within himself, however. Throughout the story, he is thrilled by the 'raw energy' of fossil-fuelled 'metal beasts' (pp. 30, 12) and by the 'fierce and wayward beauty' of the industrial landscape of power stations (p. 72). In the city, Halloway meets the like-minded Buckmaster (the name perhaps an ironic nod to eco-visionary Richard Buckminster Fuller), whose 'limitless appetite for steel, power, concrete and raw materials' is contrasted with the 'self-denying, defeatist lives of the engineers and architects' back home in the eco-community (p. 45). For Halloway, and presumably also for Ballard, the sustainable ecotopia offers none of the vitality of modernity. The narrative kinetics of solar only come into play when the technology malfunctions and the reader is reminded, unavoidably, of its materiality. Hydrocarbons fashioned and eventually wrecked the industrialized world, so solar energy created and destroyed Halloway's. Both energy sources seemed to be locked in grim competition. Ballard tries to have it both ways, with both fossil fuel consumption and the culture of renewables in his satirical sights.

Ballard's environmentalist satire seems to be directed at not only the 'garden city' projects of twentieth-century Britain which had produced carefully planned yet arguably soulless environments but also those 'alternative' and 'appropriate' technology movements and eco-communities emerging in the early 1970s, whose main aims were 'self-reliance, decentralisation, simplicity, minimum environmental impact, renewable resources'.[19] Inspired partly by Stewart Brand's Whole Earth Catalog collective and its associated American communes, the British communities included the National Centre for the Development of Alternative Technology (now known as the Centre for Alternative Technology, or CAT) in Machynlleth, mid-Wales.[20] CAT had been established by the businessman Gerard Morgan-Grenville (1931–2009) in 1973, in the disused Llwyngwern slate quarry.[21] It soon became a visitor and education centre promoting sustainable living in its many aspects, and continues to this day. In an interview with *The Sunday Telegraph*, Morgan-Grenville explained that his scientific and social experiment was attempting to show that current standards of living can be maintained without 'indulg[ing] in technologies which are potentially hazardous' or adopting a 'rapacious attitude to resources'.[22] While the 1973 energy crisis gave some credibility to such ventures, and Britain was becoming a place of considerable research in solar power, CAT's residents could still find themselves derided as 'nut cases' for their 'obsession' with renewables and recycling.[23] Ballard seems to capitalize satirically on this public perception and, yet, despite the thrill attached to the combustion engine and neon lights of modernity, it is Halloway who is left both deluded and obsessed, seeking another city to revive. Moreover, the plot conforms to the narrative of ecological collapse caused by unbridled consumption which CAT and other groups put forward at this time, through campaigns against 'growth mania'.[24] The story, therefore, offers a relatively nuanced sense of the psychology of energy transition, recognizing not only the visceral kick and narrative nourishment of hydrocarbon society (and its inevitable destruction) but also the necessity and, at least to some, the apparently boring worthiness of solar power and the eco-communities in which its usage is prevalent.[25] Perhaps, as Ballard suggests, our future will be one of apparently dull sustainability, degrowth and nostalgia for industrial grime. But better that than no future at all.

The psychology of energy is one of the key challenges to the transition to renewables. As Naomi Klein argues, the solution to global warming is to 'fix ourselves', to change the way we think about our lives and lifestyles, in order to make social transformations.[26] How do we counter the cultural attractiveness of the internal combustion engine? Perhaps dullness is the inherent 'cultural logic' (to use Peter Hitchcock's term) of solar energy?[27] Or should we think about solar more heteroglossically, since it can be a death ray too? Arguably, solar energy only seems to become exciting within narratives if it is an agent of destruction, aided of course by human (or alien) intervention, both intentional and accidental.[28] If this is the case, we might wonder then how climate campaigners could harness a more positive version of the solar technological sublime: one which creates, not destroys, and which avoids anthropocentric ideas of human dominion and the reduction of complex environmental phenomena to 'natural resources'. While Ballard's is a distinct, singular view, some authors since – including at times when the transition to renewables has looked much more assured – have also struggled to imbue solar with a cultural and narrational vitality. As we will see, though, others have attempted to overcome this failure of the utopian imagination.

We remain on the geographical periphery in Norman Rush's *Mating* (1991), in some ways a mirror image of the psycho-thermodynamics at play in 'Ultimate City'. In this novel set in the early 1980s, Nelson Denoon is an anthropologist who establishes in the Botswanan village of Tsau a utopian community ('Sekopololo') fuelled by solar energy. Near the beginning of the novel, Denoon (whose name perhaps denotes the Sun's daily zenith) proclaims the political credentials of this power when cornered at a meeting of Western expats. For Denoon, using the 'free energy of the sun' for 'every mechanical process without exception' augurs a new kind of government: neither left nor right, but a 'solar democracy' built upon mechanization of labour and changes in land usage, and characterized by political and social liberty, sexual equality, economic prosperity, and scientific and cultural enlightenment. Villages could capitalize upon their decentralized energy infrastructure, and become (oxymoronically) 'engines of rest' (p. 86), with their inhabitants 'a better thing than rich: they could be free'.[29] The response from his expatriate audience is muted at best, however, if not downright critical, and the rest of the novel is in part an exploration of whether Denoon can realize this dream.

We view Denoon through the eyes of the anonymous female narrator, who becomes his lover, but is wary of his *'hubris about solar technology'* (p. 88). Given the technological determinism in Denoon's political visions, it is no wonder the narrator sees these revolutionary social aspirations as 'Promethean' (p. 364). Appropriately, Denoon is a fan of William Blake and in possession of a kind of 'demonic energy' (pp. 277, 261), with his lifelong 'vitromania' (p. 155) symbolic of his personal and political investment in his utopian project, and hinting at its fragility. The village itself is a beacon of illumination, covered in a 'profusion of glints and flashes of reflected light' from 'various mirrors and solar instruments and other glass oddments' (pp. 152, 167). The reality of solar-powered life fails to match Denoon's ambitions, however, with the residents complaining 'continually' about 'having to keep adjusting the tracking mirrors' for their solar ovens (p. 190).

The novel's parallels with Joseph Conrad's *Heart of Darkness* (1899) are readily discernible. Like Kurtz, Denoon has a powerful voice and charisma, and an intention to bring both metaphorical and literal enlightenment to the so-called 'dark continent'. Fascinated by their initial encounter at the expat meeting, the unnamed narrator goes on a Marlow-like quest to find him. Denoon's neocolonialism, however, originates in a more recent, liberal intellectual – and less Eurocentric – tradition, and he is perhaps partly inspired by Rush (b.1933) himself, who had served as co-director of the Peace Corps in Botswana from 1978 to 1983, and in 1950 had been incarcerated for nine months for his anti-war activities.[30] The ironic lens through which Denoon is treated, therefore, forms part of Rush's self-parody of utopian ambition. We might also wonder whether there is something of another American supporter of renewable energy about Denoon. President Jimmy Carter (b.1924) had seen the energy crisis as the United States's wake-up call, and in a series of speeches and initiatives he advocated a national transition towards renewable, clean forms of energy.[31] On 20 June 1979, he delivered a press conference on the roof of the White House, where thirty-two solar water-heating panels had been installed. His central message was one of empowerment and self-sufficiency: 'No one can ever embargo the sun or interrupt its delivery to us'.[32] For some environmental and energy activists, Carter had not gone far enough, particularly in appearing to blame individual consumers rather than introducing broad structural changes. Nevertheless, Ronald Reagan's 1980 presidential election campaign sought to capitalize on the incumbent's discourse of fossil fuel scarcity by attaching himself to cowboy iconography which emphasized control of an abundant natural world.[33] Like Denoon, Carter was subsequently dumped by those he sought to serve.

Although Denoon is in many ways a caricature of a solar enthusiast, his active characterization (including his sexual vitality) at least counters the perceived dullness of 1970s eco-champions. But the treatment of solar power becomes a victim of the narrative's need to parallel Kurtz's turn to savagery with Denoon's thwarted 'scientific utopian[ism]' (p. 89), and the rejection of solar technologies by the 'native' Botswanans raises questions about the novel's representation of racial difference. Both 'The Ultimate City' and *Mating* present characters whose identities, psychologies and politics are bound up deeply with their experiences of energy. They remind us that our energy behaviours are not based merely in economic or social logic, or determined by the technology itself, but are also subject to individual thoughts, desires and ideologies. Such factors might seem to be on the periphery, but are central to our relationships with forms of energy, and will shape their future.

Celling the Sun

New forms of energy – both real and fictional – are particularly attractive to the popular imagination. Proffered as means of solution and salvation, they signify futures of limitless power and infinite possibility. But the newness of energy forms and schemes also makes them acutely vulnerable to fraud of various kinds, as promises are made which it cannot be verified will be fulfilled. To the contemporary writer, solar power has, therefore, proven to be a fertile theme through which to explore matters of intellectual property, the value of scientific discovery, notions of scientific truth, public trust in science and the socio-economic pressures science faces within society. In both Stephen Poliakoff's play *Blinded by the Sun* (1996) and Ian McEwan's novel *Solar* (2010), artificial photosynthesis – a technological 'faking' of nature, sometimes called 'biomimicry'[34] – is a trope used to explore fraudulent or embellished science, and the modern research centre or laboratory becomes a meta-textual space of discovery on the boundary between fact and fiction, truth and hyperbole.

Set in a declining chemistry laboratory at a British university, the plot of *Blinded by the Sun* revolves around a 'Sun Battery' purported to split water photocatalytically:

Find the right chemical to act as a catalyst – shine a light, a beam, above all the sun – and you can create hydrogen out of sunlight and water. Hydrogen, which will run in planes, cars, anything you want. And when you burn it, it will turn back to water. Polluting nothing.[35]

Yet all is not as it seems. The inventor of the Sun Battery is Christopher, who (claiming to be protecting his intellectual property) refuses to divulge the secret of the process even to Al, his line manager. Al, the focus of the play, is dazzled by the prospect of the accolades and funding which he believes will come the laboratory's way, securing its future at a time of uncertainty in UK academia (p. 34). Even Al's seven-year-old daughter is infected by this enthusiasm, drawing a picture of '*a battery, like a colossal domestic torch battery, warming up the world, with the sun coming out of it*' which rescues the 'rain forest … and all the animals' (p. 39). Rather than publishing his research, however, Christopher announces it via a press conference, which Poliakoff (b.1952) admits was inspired by the public claim of 'cold fusion' made in 1989 by Martin Fleischmann and Stanley Pons of the University of Utah.[36] Unfortunately, Al's dreams are crushed when he figures out that Christopher has faked the photocatalytic process using calcium hydride, made from baking powder and bleach: 'An idea that sounds so pure, sunlight and water, is a fucking con. He's used baking powder, for Chrissake! It's as crude as that' (p. 62). Al's anger, of course, is also directed at himself, as he realizes that his own ambition has willed this discovery into existence. Verbal and dramatic irony is important to the play, particularly when Al (as yet unaware of the fraud) explains to a visitor that 'when you see [the battery] the enormity of it all will not be obvious. It'll just look like a tedious little tube' (p. 31), or admits that the lab notes he has been allowed to read are 'a little cloudy' (p. 41).

Advised against it by his colleague Elinor, Al becomes whistle-blower and the fraud is made public. Remarkably, Christopher seems to do rather well out of the situation, with daily invitations to speak (p. 87), as Poliakoff satirizes what he sees as a disintegration of integrity within the culture of science, as 'showbiz and science are getting ever closer' (p. 99). Christopher never admits his deception, though, and continues to claim that he 'just couldn't repeat the experiment, that's all' (p. 88). Al does best of all out of the affair, becoming a radio personality and bestselling author, embodying Poliakoff's lament that this is now a world where 'marketing is the key, not creativity' and administrators, not true intellectuals and innovators, can flourish.[37] Underlining this professional shift, at the end of the play Al closes down the laboratory, and Elinor, the 'true' scientist, is made redundant. Playing with ideas of appearance and reality, celebrity and integrity, scientific and public value, *Blinded* uses solar technology as a metaphor for the struggle to maintain science as a beacon of pure white light – an idea taken much further in McEwan's *Solar*.

Near the beginning of *Solar*, the protagonist Michael Beard confesses that he considers the term 'solar energy' to have a 'dubious halo of meaning, an invocation of New Age Druids'.[38] 'Halo' is a pertinent metaphor, exploiting the word's own multiple denotations: originally scientific (a circle of light around a celestial body), and later religious ('glory' or 'nimbus'). The word 'energy' is itself polysemous, normally pertaining to the power possessed and exerted by a body, and in science specifically to the capacity to do work, with 'power' as the rate at which this work is performed.[39] In English, it originally referred to strength of expression in speech or writing and, as we have witnessed many times before, solar energy lends narratives a unique metaphorical force. Beard recognizes that although manifestations of solar energy can be isolated scientifically across spectral wavelengths and also defined according to the technologies that capture sunlight, in discourse the term somewhat suspiciously produces further inflections related to an ethereal, primal symbolism. Like many fictional scientists before him, the rationalist Beard comes to learn that the symbolic power of the Sun and its technologies cannot be dismissed so easily.

Beard is a world-famous physicist whose adaptation of Einstein's theory of light quanta (the 'Beard-Einstein Conflation') rewarded him with a Nobel Prize, but who is now in the twilight of his career and on his fifth failed marriage. We meet Beard in 2000 as Head of the new 'National Centre for Renewable Energy' in Reading, not 'wholly sceptical about climate change' but 'unimpressed by some of the wild commentary' associated with it (p. 15). (The UK's real-life National Renewable Energy Centre (NaREC) was established in 2002 in Northumberland, but many aspects of its business have since been privatized.)[40] The (fictional) centre's first main project is to roll-out domestic micro wind-turbines, which are often cited as a waste of time and money.[41] Meanwhile, one of the junior researchers, Tom Aldous, is telling the uninterested Beard relentlessly that the Centre should be investing in solar research. It is not until Beard gets hold of Aldous's papers (in tragicomic circumstances) that he sees their value. Beard then leaves the centre to pioneer a form of artificial photosynthesis, seeing the man-made mess of the world as the means of his redemption professionally and personally, with the Sun sending its 'photon torrent to illuminate and elevate his labours' of salvation (p. 184) – the mere action of light figured as a creative force.

Solar does not go into the kind of detail about artificial photosynthesis that McEwan's *Saturday* (2005) pursued in relation to neurosurgery, perhaps because it was then a relatively new and unpublished area of research, and it has been argued that the novel's 'depiction of futuristic renewable energy is incidental to Beard's conspicuous embodiment of self-deception and gluttony'.[42] Nevertheless, it is hinted that the process Aldous invents (and Beard takes forward) is a type of 'nano-solar' (p. 25), which involves the 'nanostructuring' of the semiconductors: one situational irony of this being that the rotund Beard is seeking to create thinner and thinner cells.[43] This sense of the tiny resonates throughout the novel as a narratorial and figurative strategy. It is reflected in the small-scale, small-man *nano-narrative* of the corpulent and self-deluded Beard as a synecdoche and symbol of humanity's abuse of ourselves and the planet. In this 'nano-apocalypse', the little details and detritus in Beard's life and career are revealed to have bigger implications which eventually come back to burn him, despite his plan of 'stick[ing] to photons – no resting mass, no charge, no controversy on the human scale' (p. 144).

Much of the novel's media coverage attended to the comic and satiric modes McEwan (b.1948) sustains in *Solar*, a relative departure for him. McEwan's decision perhaps tells us something about his view of art's agency in the contemporary climate change debate: it can only vex, not divert, despite some ecocritics hoping this ostensibly 'cli-fi' novel (which emerged from McEwan's visit to the Arctic as part of the Cape Farewell 2005 Art/Science Expedition) would be a 'pivotal' influence on public opinion.[44] The power of literature per se is given a more positive treatment in the novel itself, however, when we discover that at university Beard read Milton to impress a girl. This sublime, illuminant author who had prayed for inspiration from the 'Celestial light' came to partly stimulate Beard's optical research (p. 201), this unintentional effect prefiguring the scientist's late, accidental 'conversion' to solar.[45]

McEwan perhaps also recognizes that there is something inherently ironic about solar energy projects as part of a solution to climate change: harnessing the Sun artificially as a way of slowing down the escalating *uncontrolled* gaseous capture of the Sun's infrared energy. This cosmic irony is encapsulated in the exchange between Beard and his business partner and project manager, Toby Hammer. Amid the growing number of climate change deniers and the potential economic collapse of their scheme, Hammer fears that 'if the place isn't hotting up, we're fucked', to which Beard offers the reassuring: 'It's a catastrophe. Relax!' (pp. 216-17). Renewable energy is targeted satirically as a sector attracting a new kind of financial and technological speculation inspired by a variety of motives (and more likely to be venture capitalism than environmentalism), and acutely vulnerable to fraud, both economic and intellectual, because of its novelty. This has been borne out in reality, in some ways, with a number of high-profile scandals involving the roll-out of renewables.[46]

In plagiarizing his dead employee's work, Beard is a kind of comic Prometheus (that first industrial spy) – a stealer of knowledge regarding the Sun's power. There is also something tentatively Promethean about his aspirations. As solar speculators, both Aldous and then gradually Beard become overawed by the sublimity of the Sun, the simplicity of the concept of emulating nature, and the purity of the energy source, and forget the practical difficulties and commercial realities. In effect, Beard comes to believe his own PR, developing the self-regard to conceive that 'the world would be saved' by his showcase scheme (p. 226), and he falls prey to the third-person narrator's irony:

> The day after tomorrow a new chapter would begin in the history of industrial civilisation, and the earth's future would be assured. The sun would shine on an empty patch of land in the boot heel of south-west New Mexico ... the Nobel laureate would throw a switch and the new era would commence. (p. 213)

The novel is not wholly critical of Beard's pride, though, in that his pomposity provides the inspirational energy for the project to be taken forward – it gets things moving, even if not for the right reasons initially. This is not to say, however, that the novel itself 'holds out the hope that science will solve the crisis [by providing] new technology', or celebrates the genius of Beard.[47] The novel acknowledges that to allow such technology to flourish, humanity also needs to acknowledge its appetites and the problems they cause, and deal with them accordingly. It also raises the question of whether protecting intellectual property rights really matters when confronted by climate crisis.

The excruciating moment of Beard's comic hubris comes when he takes a panoramic view of his nano-solar installation in the New Mexico desert:

> The twenty-three big tilted panels had a dull gleam under the ferocious sun. They were fed by a mess of pipework and valves. … Power lines on poles led to the nearest of the ancient wooden pylons that tottered in succession across the immensity of semi-desert. … What was new were the hundreds of people … moving importantly between tasks, …
>
> Beard said, 'Toby, you're a genius!'
>
> Hammer nodded in grave acknowledgement. 'I like to bring things and people together. But, Michael. This is your invention. The genius is you.'
>
> Feeling serene now, Beard nodded in return. This was how friendship should be. (pp. 241–42)

For all of Beard's initial rejection of the symbolic power of solar energy (as something associated with the Druids), he is himself taken up within the ritualistic veneration of a project that is filling with new meaning a sublime landscape already associated with pure heat (a sun-trap bringing life and hope rather than death). The technological and natural sublime combine to exalt both inventor and project director in a mutual back-slapping, bromantic moment. It is an instant tinged with especial irony, given both Beard's intellectual thievery and the knowledge that the planned 400 acres of land for the plant had been reduced to 25, due to a reduction in credit.[48] This credit problem meant that 'worst of all, because they were the core and symbol of the project, a mere twenty-three panels tilted skywards instead of one hundred and twenty-five' (p. 212). Beard's paltry offering is, therefore, a diminutive caricature of the kinds of desert solar projects operational at around this time, many of them involving immense concentrated solar power (CSP) towers which heat a thick fluid (typically molten salt) to drive steam turbines and hence generate electricity (Figure 16). Nevertheless, the exchange between Beard and Hammer highlights an important point about the operation of the technological sublime within the climate change debate. Might our veneration of science induce a sublime response to images of enormous CSP towers or vast arrays of solar cells, with the potential to hinder the green agenda by cultivating the public perception that human reason (itself symbolized in these technologies) will automatically find a way out of the impending problems of energy shortage and climate change? For a start, the 'functionalist aesthetics' and/or 'ethical sublime' of such structures

Figure 16 Concentrated solar power tower. Adobe Images.

obscure the problems of long-distance transmission, storage and habitat damage, they raise.[49]

Moreover, these solar-sublime technological solutions encourage us to focus on changing production, not reducing our consumption. One potential future in this context is offered in *Blade Runner* [2049] (2017), which begins with techno-pastoral, aerial views of vast solar thermal farms of CSP towers and heliostats in the Californian desert. Although the film never explains this, it seems that these power plants have been rendered inactive by the dense atmospheric pollution, and it is unclear whether they have ever been operational.[50] John Gerrard's digital art installation 'Solar Reserve (Tonopah, Nevada) 2014' uses CSP to make a similar point. The computer-generated piece depicts a solar thermal power tower, surrounded by 10,000 heliostats, in the Nevadan desert, simulating the movement of the Sun over the course of the year, and the corresponding movement of the mirrors. Every 60 minutes, the perspective on the plant shifts from ground level to an aerial view. Gerrard (b.1974) himself called the work 'an apparition, existing solely as software code', and reads it metaphorically: 'To what degree is the optimism of renewable energy itself an apparition if we do not reduce global energy consumption?'[51] The solar-sublime is perhaps entirely notional, its aesthetic experiences gesturing towards a phantom vision of the future that might not come to pass.

McEwan's *Solar* reinforces that it seems reckless to leave our fate in the hands of small-measure government intervention, market forces and particularly scientific 'genius'. As the environmental author Kenneth Brower writes: 'The notion that science will save us is ... the sedative that allows civilization to march so steadfastly toward environmental catastrophe. It forestalls the real solution, which will be in the hard, nontechnical work of changing human behavior.'[52] Whether contemporary cli-fi can mobilize environmental action is a moot point. A recent survey of cli-fi readers suggests that such literature is effective at enabling readers to 'imagine potential futures', but it is as yet unknown if this translates into 'meaningful changes' in behaviour or politics.[53] Rather than giving us an apocalyptic parable or visionary ecotopia that could have been overbearing in its morality and earnestness, McEwan instead wrote a satirical novel that counteracts the many images of the single, heroic scientist we have encountered. *Solar* shows us that one scientist, or one inventor, or one writer, cannot change the world on their own, and that we must all work together to solve what is perhaps our biggest problem.

Decentralizing the Sun

Beard's project, of course, is relatively modest in comparison to some of the grand technological solutions to climate change proposed by real scientists. 'Solar Radiation Management', for example, involving deflecting space mirrors, spraying chemicals into the atmosphere or other geoengineering solutions, gets a short shrift from environmental writers. Klein points out that the sulphate aerosol solution could potentially make blue skies a thing of the past, and weaken solar influx so as to 'reduce the capacity of solar power generators to produce energy (irony alert)' (*This Changes Everything*, p. 259). Jeremy Leggett asks: 'Why do scientists fly off on such risky or even crazy tangents when the well of real-life renewable-energy, energy-storage and energy-efficiency alternatives is so deep and productive?' He posits two reasons. One is to do with scientific specialism: an expert on space technology (for example) pursues a solution within their own field, even if better alternatives within other (terrestrial) fields exist. The second is about the individualism rewarded by institutional science, a culture of 'lionizing the stars'. And if such a scientist is 'a famous exponent ... maybe even a Nobel Laureate, [they] are actually in danger of being taken seriously'.[54] Whether Leggett's analysis is true or not, it is certainly similar to the scenario that plays out in McEwan's novel. However, Leggett does not take account of another factor, one that we have seen at work throughout the centuries in relation to solar devices: the affective thrill of the technological sublime. This sublimity comes from not only its immediate but also its 'creational' aesthetics that suggest the technical mastery of its creator. But how might we encourage a reorientation of our values in relation to technological solutions? We could perhaps do worse than by turning to poetry.

The move to renewables involves a shift away from the state- or corporate-run, grid-based energy supply (and its technologically sublime infrastructures), and towards the individual producer/consumer: their activities now blurring into one, as signified by the slightly clumsy portmanteau word 'prosumption'.[55] Anyone who owns a roof can buy or lease photovoltaic panels if the price is right (and the mortgage lender permits them), and social enterprise projects allow non-home owners to tap into community-run micro-grids. The psycho-thermodynamics of this shift to 'distributed generation' are already being studied. Installing PV on a domestic home changes the consumer's relationship with electricity production, since it increases the visibility of the resources. Off-grid, they become more aware of what they consume, and consequently more likely to conserve energy and pursue further efficiencies; they become appreciative of the source and of the technology which has harnessed it.[56] Microgeneration via community-owned systems, meanwhile, is perceived to be the important factor in national transitions to sustainable energy.[57]

The idea of the individual producer/consumer is at the heart of Derek Mahon's (b.1941) 'Its Radiant Energies', the first in a sequence of nine ecologically themed lyrics entitled 'Homage to Gaia' (2008). The poem explores 'how to live / in the post-petroleum age', at times offering practical advice like a popular guidebook to sustainable lifestyles.[58] One must 'run the house with clean / photoelectric frames', their cleanliness both physical and ethical, in order to 'focus' the Sun's 'radiant energies'. This 'focus' is metaphorical rather than literal, hinting that these 'energies' are not only physical but psychological or spiritual too (indeed, of all the previous solar harvesters we have encountered, it is John Dee who seems most echoed here). The poem revels in the juxtaposition of the cosmic and domestic, the local and global, the spiritual and technical. The solar panels have an elegant, rational 'composure', a 'heliotropic quiet', as their 'light-drinking polysilicon / raises its many faces' to the 'Great sun' which, in contrast, is 'roaring' far above. It is the panels' ordinariness, their very anti-sublimity, that makes them so appealing, especially since the Sun is 'far from our atmosphere' and yet has immense effects upon it. This line in particular seems to embrace the cosmic irony we have observed elsewhere: that the Sun's energy both contributes to and offers potential salvation from climate change.[59] Therefore, while there is an inherent verticality in the poem, as it constantly looks upwards to the Sun in a kind of reverence, the last line appeals to both Sun and reader to 'Remember life on Earth!'. Salvation will not come from the Sun alone, but from appreciating 'its radiant energies' appropriately and living in a kind of synergy with them.

More recently, however, we find the decentralized, community-driven roll-out of renewable energy being replicated in the artistic sphere. Not content with only raising awareness, the Icelandic-Danish artist Olafur Eliasson (b.1967), most famous for the sublime artificial Sun of the 'The Weather Project' (Tate Modern, 2003), co-designed the 'Little Sun' in 2012. This solar-powered lamp is marketed as 'the power of the sun in the palm of your hand', articulating the empowerment it aspires to bring to those without regular electricity supplies, particularly by facilitating night-time reading. The eco-aesthetics of its sunflower shape emphasizes the 'naturalness' of the power source, while the company's business model – of dispensing energy justice through subsidizing off-grid purchasers in Africa – reframes 'community' as global, and tries to circumvent the slow or inactive green initiatives of national governments and international corporations.[60] Each 'Little Sun' becomes part of a cumulative, world-spanning Joseph-Beuysian social sculpture, seeking to change society through everyday interventions.[61]

New kinds of connectivity have also enabled the development of *creative* sustainable communities. The SF movement of 'Solarpunk' has recently emerged, first online and now in print, notably with the *Sunvault* anthology of text and image. Solarpunk is concerned with not only ecological but also progressive imaginaries more generally, to create 'a more optimistic future in a more just world'.[62] According to *Sunvault*'s editors, Phoebe Wagner and Brontë

Christopher Wieland, this utopian genre 'emphasizes innovative interaction with both our communities and our environment; socio-environmental thought and creation, rather than merely survival in a decaying world', a counterpoint to the more dystopian mode of cyberpunk.[63] Notable pieces within this anthology include Camille Meyers's transhumanist 'Solar Child', which imagines the genetic engineering of a 'photosapien' species as part of a wider 'photobio' research programme (pp. 185-94). This interest in augmented biology is also reflected in Jack Pevyhouse's poem about 'Solar Powered Giraffes' (p. 87), and in 'Solar Flare', Christine Moleski's illustration of a human enhanced with grafted electronic cells and seeming to harvest (or possibly emit) sunlight through their eyes (p. 133).

Sara Norja's contribution, 'Sunharvest Triptych', is a kind of georgic that celebrates solar panels as themselves the abundant produce of the land- and cityscape. Rather than glorifying these panels as the product of human achievement, their copiousness is rooted in their harmony with, not subjection of, their host environments:

> the fieldfuls, the rooftopfuls,
> the stained-glass-windowfuls of them:
> the harvesters of energy[64]

The poem is a 'Triptych' not only in its three-stanza form but also as a work of art generated or inspired by (solar) panels. At the poem's Milton-like climax, the speaker recognizes the personal, pseudo-spiritual value of the Sun's power, figuring herself as a panel that draws out the internal light:

> In the half-dark
> I harvest the sunlight
> lying in wait, after all,
> within me. (p. 220; ll. 70-73)

Reminiscent of *Paradise Lost*'s invocation to the 'celestial light', 'Sunharvest Triptych' posits poetic labour as a kind of photovoltaics, with the speaker's ethical energy existence providing a nightly renewal of the self's creativity. The poem tries to persuade us that solar power makes not only ethical and economic but also artistic sense. Who knows what the agency of such an idea might be? Only time will tell. But by acknowledging and reflecting upon the sunlight within, we can begin to imagine how to renew ourselves and the world around us.

After Sun

On the sixtieth anniversary of the UN Energy Commission, the president of the Technological Institute of the Federation of Southeast Asian Republics, Lee Ngo Nhu, addressed the UN General Assembly to remind them of the great strides in solar energy which had allowed humanity to pursue 'spiritual fulfilment'.[65] The speech celebrated how the Odeillo solar furnace developed in the 1970s had shown that solar power could obtain temperatures higher than necessary for the 'thermal "cracking" of water to produce hydrogen' (in other words, photocatalytic dissociation of water), and that the technology necessary to store, transmit and utilize hydrogen safely had also been available since that time (p. 87). Fortunately, this research proved vital when the 'Great Energy Crisis' hit, and the special energy conference in Prague in May 2018 agreed that resources would be pooled between the industrialized nations, reversing the many years of underfunding on 'alternative technologies of energy' (pp. 83, 87). A 'solar-hydrogen system' was then developed on a global scale, with millions of kilometres of natural gas pipelines converted to handle this new source of energy, and a 'cheap, simple system of solar technology' was perfected by the electrochemical engineer Harold P. Lambert from the University of Wisconsin (pp. 88, 91). Unfortunately, however, 'in direct correlation with increases of CO_2 concentration in the atmosphere', the mean global temperature had already 'risen 6.5°C by 2023, and the polar icecaps were melting' (p. 90).

As you will have guessed by now, and perhaps much earlier: this speech is a fiction, imagined as being delivered on 16 November 2078. It was written by the engineer and science journalist David A. Mathisen, for the UNESCO journal *Impact of Science on Society*, and was published in their January–March 1979 issue, celebrating the centenary of Einstein's birth. Mathisen uses the form of the conference address to weave an intricate Heinlein-esque speculative future history, which includes terrorist attacks and nuclear disasters taking place in the early twenty-first century. His speaker 'Lee Ngo Nhu' explains that inexpensive hydrogen fuel cells were made commercially available by July 2029, bringing 'a new era of energy independence … for the consumer'. This 'abundance of energy' meant that the 'recycling of materials became a highly cost-effective enterprise', resulting in 'full employment everywhere', and the elimination of the 'gross maldistribution of the planet's wealth' (pp. 92, 93). Before this happened though, a group of activists called 'Citizens against Technology and Technocrats (CATT)' tried to bomb 'everything that smokes' and pronounced that scientists and engineers were 'enemies of the people' because they had created the world's hydrocarbon infrastructure. CATT – perhaps an intentionally grossly distorted caricature of the CAT alternative technology centre in Machynlleth – received 'frightening' popular support, with the police 'count[ing] on little aid from the average citizen in putting a stop to them' (p. 90). Mathisen, therefore, provides us with a techno-utopian vision of hope, but embeds within it more subversive voices to remind his scientist readers that it is they who will receive much of the blame for global warming.

One reader of Mathisen's fiction was Keith Barnham, a young physicist who stumbled across the journal in 1979 while browsing new publications in the library at CERN. Barnham 'was captivated, slumped to the library floor and drank in every word', and 'decided there and then' to change career to photovoltaics research. Barnham became a leading PV researcher, founding the company QuantaSol to investigate and commercialize the use of quantum wells within solar cells, and since retirement from Imperial College London has become a prominent public advocate for solar energy in the UK. In his popular science book *The Burning Answer* (2014), Barnham followed Mathisen in writing his own forecasts for the future of solar, but over the more modest period of fourteen years, until 2028. Barnham's book contains some unexpected formal innovations, incorporating a 'Manifesto for the Solar Revolution', and open letters to EDF (the French utility company who had been approved to build the UK's Hinkley Point C nuclear power station), the UK Labour and Liberal Democrat parties and a 'major fossil fuel company'. Barnham's own career is, therefore, testament to the power of solar stories, and it was the fictional mode Mathisen used that grabbed his attention in particular. Writing in 2014, Barnham found '2079' still 'well worth a read, given the accuracy of its predictions'.[66] However, the value of this fiction does not really lie in its prognostications, but rather in its use of the UN speech as a rhetorical form, its placement in a scholarly journal and, above all, its investment in speculative fiction as a mode that has the potential to inspire environmental action.

It is not surprising that speculative fiction has proved attractive to environmentalists and enthusiasts of renewable energy. As Ursula K. Heise has explained, the global, pseudo-epic storytelling of SF means it 'makes sense' that its motifs and structures 'keep cropping up in environmental nonfiction, books whose principal concern is to drive home the *reality* of current ecological crises'.[67] Speculative (non-)fictions of solar energy, however, combine the catastrophism often encountered in environmentalist writing with appeals to the everyday and technocratic imaginations, and they often cast solar technology itself, rather than scientists, engineers or activists, as the closest thing to the hero. Jonathan Porritt's speculative history of the future *The World We Made: Alex McKay's Story From 2050* (2013) is a case in point. Porritt (b. 1950), former director of Friends of the Earth, adopts the voice of a fifty-year-old history teacher giving fifty snapshots of the social and technological changes of the previous thirty odd years.[68] The book's blurb claims that, for the last forty years, Porritt has 'tried every conceivable way of persuading people to share his excitement at the prospect of a genuinely sustainable world. ... Finally, he discovered Alex McKay, providing a unique opportunity to connect that world in 2050 with what we can do today to help make it a reality'. The blurb is there, of course, to sell the book, but it is striking that it asserts the potential of imaginative works to provoke social action.

As one might expect, solar power has a central role in *The World We Made*, especially given that it describes a world which faces peak oil in 2017, and which suffers cyber-attacks against five US and two UK nuclear reactors in 2019 (p. 60). To Porritt's narrator McKay, the achievement of grid parity 'helped redefine the American Dream, as it freed both individuals and the country as a whole from dependence on big foreign energy companies', with the irony being that 'it was the Chinese that originally made all this possible' through solar panel manufacture (p. 15). The world, including the Global South, moves towards energy justice and a concomitant improvement in living standards, with the 'Solar Salvation' solar thermal and PV scheme established in slums outside Lagos (Nigeria) in 2023 enabling a women's cooperative to bring health care, education and employment to the area (pp. 218-19). Reminiscent of Gernsback's strategy in *Ralph 124C 41+* (1911), Porritt embeds recent real innovations as the forerunners of imagined projects, such as the 'Desertec' initiative established in 2009, which sought to supply Europe with electricity generated by large-scale CSP, PV and wind turbine plants in the Sahara (p. 16). Photographs, maps and diagrams of technologically sublime CSP facilities are included to galvanize our sense of their functionalist aesthetics (pp. 16-17, 19-21). By 2050 such projects bring 'prosperity and stability' to North African countries, although this does not happen overnight. Porritt does not shy away from speculating about some of the practical, socio-economic and political problems that would occur initially though, but with the point being that the political will was found to resolve them (p. 19).

McKay's global conclusion consolidates his overall message that science and technology created the ecological crisis but, via two recent phenomena, have now 'opened up the world to allow empathy to flourish'. The first is the internet, apparently bringing 'effortless, ubiquitous connectivity for every single one of us' by 2050. The second is the 'solar revolution', since 'sunshine isn't like [fossil fuels]: it's "owned" by the whole of humankind, to be used in the interests of every single citizen on Earth'. McKay hence argues that solar is the 'greatest ever technological leveller in the history of mankind' (p. 269). This techno-utopian rhetoric negates, of course, the potential for energy injustice and inequality that the global roll-out of renewables might actually entail (which also has a bearing upon access to telecommunications). Yes, sunshine itself (like the internet) cannot be owned by one person, but the technology which harnesses its power can be restricted or monopolized – as Arbuthnot's 'Humble Petitioner' feared back in 1716.

Still, there are reasons to be cautiously optimistic about the future of energy. As Porritt predicted, in lots of countries grid parity has been reached, and the levelized cost of electricity (LCOE) generated by renewables is now (in 2019) equivalent to and, in many cases, much lower than fossil fuels.[69] One of the principal challenges to full transition is the intermittency of supply, with solar power output at its highest in the middle of the day and wind power normally late at night – both times of low energy demand. Hence the need for efficient storage which it is expected will be supplied by lithium-ion batteries (or those manufactured from more abundant alternatives), used particularly in electric vehicles. Indeed, the convergence of renewable energy and electric vehicles may be the key to solving the intermittency problem, with vehicle batteries being charged at times of peak output. All this has happened very recently, at a massive pace. Yet it is unlikely that reductions in cost, and the availability of electric cars and associated infrastructure, will transform our energy behaviours entirely. And time is not on our side.

Therefore, something else is needed, now, at this moment of climate crisis. According to Bill McKibben (b.1960), probably America's most famous environmentalist of the twenty-first century, there are 'two relatively new inventions that could prove decisive to solving global warming before it destroys the planet. One is the solar panel, and the other is the nonviolent movement'.[70] Again, solar technology is proffered as a means of salvation for our increasingly fragile planetary ecosystem, but McKibben recognizes that technology alone cannot decarbonize our society and that it will also involve global efforts to change both government policies and the ways we collectively choose to live our lives. This 'nonviolent movement' (in actuality, far from a 'relatively new invention') cannot just be about protest: it needs to offer realistic visions of a sustainable future that individuals, corporations and policymakers can believe in and work towards collectively. And this is where acts of the imagination, and the mobilization of history, could contribute. To realign our behaviours must involve looking at the discourses, narratives and images we have used to represent and imagine the production and usage of energy. While solar power, in its many forms, has (paradoxically) been lauded as a novelty which will be the resource of the future, the chapters here tell us of a long history. Over hundreds of years, the cultural meanings associated with capturing solar radiance have had a significant influence upon our conceptions of the role of the scientist in society and of human agency in relation to the natural environment – and those meanings themselves could be harnessed and reoriented within the global consciousness to help create a more sustainable future. Human aversion to change is cited as another barrier to transition,[71] but might this be countered by recognizing that embracing renewables would constitute an energy renaissance, not a revolution?

Moreover, if anything should convince us that the humanities and sciences need to (return to) work together, it is our present climate crisis. For technologies and behaviours to change, there also must be a transition in how we see our disciplines of knowledge-making in relation to one another. A single intervention will not change the world to the extent we need, despite the many fantasies we have encountered which purport to the contrary. To end humanity's extractive and polluting energy activities and materialist over-consumption will inevitably generate another multifarious chapter in the history of our species' complex relationship with the Sun. Like those before, it will be one full of thought, experiment and, above all, imagination.

Notes

Introduction: Bringing the Sun into focus

1   John Arbuthnot, *To the Right Honourable the Mayor and Aldermen of the City of London: The Humble Petition of the Colliers, Cooks, Cook-Maids, Blacksmiths, Jack-makers, Brasiers, and Others* (London, 1716), p. 1; *Miscellanies in Prose and Verse by Pope, Swift and Gay* (1727–32), ed. Alexander Pettit, 4 vols. (London: Pickering & Chatto, 2002), IV, pp. 72–8.

2   'by Dr Arbuthnot' is handwritten on the British Library copy (BL 816.m.19).

3   Georgia Brown, 'British Power Generation Achieves First Ever Coal-Free Day', *The Guardian*, 22 April 2017, accessed via <https://www.theguardian.com/environment/2017/apr/21/britain-set-for-first-coal-free-day-since-the-industrial-revolution>.

4   See Robert S. Emmett and David E. Nye, *The Environmental Humanities: A Critical Introduction* (Cambridge, MA: The MIT Press, 2017), p. 55.

5   See esp. Timothy Mitchell, *Carbon Democracy: Political Power in the Age of Oil* (London: Verso, 2011).

6   See, for instance, Ross Barrett and Daniel Worden, eds., *Oil Culture* (Minneapolis, MN: University of Minnesota Press, 2014).

7   Jimmy Carter, 'Address to the Nation on Energy', 18 April 1977, accessed via <https://www.presidency.ucsb.edu/documents/address-the-nation-energy>.

8   On the naming of the Anthropocene, see Paul Crutzen, 'Geology of Mankind', *Nature*, 415: 6867 (2002), 23.

9   See esp. Sheena Wilson, Adam Carlson and Imre Szeman, eds., *Petrocultures: Oil, Politics, Culture* (Montreal: McGill-Queen's University Press, 2017).

10   Daniel Behrman, *Solar Energy: The Awakening Science* (London: Routledge & Kegan Paul, 1979), p. 261. See Chapter 7.

11   On the latter name, see James Clerk Maxwell, 'A Dynamical Theory of the Electromagnetic Field', *Philosophical Transactions*, 155 (1865), 459–512 (p. 466).

12   See 'renewable, *adj.* and *n.*', *OED Online*, accessed via <http://dictionary.oed.com>.

13   See Timothy Morton, *Hyperobjects: Philosophy and Ecology after the End of the World* (Minneapolis: University of Minnesota Press, 2013), pp. 1–2.

14 See Lucie Green, *15 Million Degrees: A Journey to the Centre of the Sun* (London: Viking, 2016), pp. 1–3, 20.

15 On explicit and implicit energy narratives, see David E. Nye, 'Energy Narratives', *American Studies in Scandinavia*, 25 (1993), 73–91.

16 David E. Nye, *Narratives and Spaces: Technology and the Construction of American Culture* (New York: Columbia University Press, 1997), p. 88.

17 Terry Macalister and Eleanor Cross, 'BP Rebrands on a Global Scale', *The Guardian*, 25 July 2000, accessed via <https://www.theguardian.com/business/20 00/jul/25/bp>.

18 See esp. David E. Nye, *American Technological Sublime* (Cambridge, MA: MIT Press, 1994), p. 62.

19 Plato, *The Republic*, trans. Robin Waterfield (Oxford: Oxford University Press, 1994), pp. 264–5 (532a-d).

20 See Ian Woodward, *Understanding Material Culture* (London: SAGE, 2007), p. 94.

21 On technopolitics, see esp. Gabrielle Hecht, *The Radiance of France: Nuclear Power and National Identity after World War II* (Cambridge, MA: MIT Press, 1998), p. 15.

Chapter 1

1 Sigmund Freud, *The Interpretation of Dreams*, trans. James Strachey, ed. Angela Richards (1953; London: Penguin, 1991), p. 252n2. Cf. James Joyce, *Ulysses*, ed. Danis Rose (London: Picador, 1997), p. 360.

2 A notable exception is the apparently successful attempt of the Greek engineer Ioannis Sakkas in 1973. See Thomas W. Africa, 'Archimedes through the Looking-Glass', *The Classical World*, 68: 5 (February 1975), 305–8 (p. 305).

3 Heinrich van Etten, *Mathematicall Recreations* (London, 1633), p. 130.

4 Diodorus of Sicily, *History* 26.18, in *Greek and Roman Technology: A Sourcebook*, eds. John W. Humphrey, John P. Oleson and Andrew N. Sherwood (London: Routledge, 1998), pp. 541–2.

5 Polybius 8.5-7; Livy 24.34; Plutarch, *Life of Marcellus* 14-17. See D. L. Simms, 'Archimedes and the Burning Mirrors of Syracuse', *Technology and Culture*, 18: 1 (January 1977), 1–24 (p. 3).

6 Lucian, *Hippias*, in *Lucian of Samosata*, trans. A. M. Marman (Cambridge, MA: Harvard University Press, 2014), p. 37; D. L. Simms, 'Galen on Archimedes: Burning Mirror or Burning Pitch?', *Technology and Culture*, 32: 1 (January 1991), 91–6 (p. 91).

7   See R. Rashed, 'A Pioneer in Anaclastics: Ibn Sahl on Burning Mirrors and Lenses', *Isis*, 81 (1990), 464–91 (p. 468), and G. J. Toomer, trans. and ed., *Diocles on Burning Mirrors: The Arabic Translation of the Lost Greek Original* (Berlin: Springer-Verlag, 1976), p. 22.

8   Robert Record, *The Pathway to Knowledg, Containing the First Principles of Geometrie* (London, 1551), n.p. See also Thomas Digges, 'The Preface to the Reader', in Leonard Digges, *A Geometrical Practise, Named Pantometria* (London, 1571), sig. A3^v.

9   See Paula Findlen, *Possessing Nature: Museums, Collecting, and Scientific Culture in Early Modern Italy* (Berkeley: University of California Press, 1994), pp. 328–30.

10   See Sven Dupré, 'Optic, Picture and Evidence: Leonardo's Drawings of Mirrors and Machinery', *Early Science and Medicine*, 10: 2 (2005), 211–36.

11   Jean Paul Richter, ed., *The Literary Works of Leonardo da Vinci*, 3rd edn, 2 vols (1883; London: Phaidon, 1970), I, 121.

12   See William Eamon, *Science and the Secrets of Nature: Books of Secrets in Medieval and Early Modern Culture* (Princeton, NJ: Princeton University Press, 1994), pp. 9, 11.

13   Giambattista della Porta, *Natural Magick*, trans. anon (London, 1658), p. 355.

14   See, for instance, Thomas Nashe, *The Unfortunate Traveller and Other Works*, ed. J. B. Steane (London: Penguin, 1972), p. 309.

15   See Eileen Adair Reeves, *Galileo's Glassworks: The Telescope and the Mirror* (Cambridge, MA: Harvard University Press, 2008), p. 120.

16   Andrea Pozzo, *The Glory of Saint Ignatius* (Rome: Sant'Ignazio, 1691–4).

17   See Ingrid D. Rowland, 'Athanasius Kircher, Giordano Bruno, and the Panspermia of the Infinite Universe', in *Athanasius Kircher: The Last Man Who Knew Everything*, ed. Paula Findlen (New York: Routledge, 2004), pp. 191–205.

18   Athanasius Kircher, *Ars Magna Lucis et Umbræ* (Rome, 1646), pp. 874–88.

19   Kircher was soundly beaten to this discovery, however, by Diocles's *On Burning Mirrors* (196–180 BC): see Benjamin Goldberg, *The Mirror and Man* (Charlottesville, VA: University Press of Virginia, 1985), p. 106.

20   See William B. Ashworth, 'Light of Reason, Light of Nature. Catholic and Protestant Metaphors of Scientific Knowledge', *Science in Context*, 3: 1 (March 1989), 89–107 (pp. 94–7).

21   See Plato, *Timaeus* 45b-c.

22   See esp. Marina Warner, *Phantasmagoria: Spirit Visions, Metaphors, and Media into the Twenty-First Century* (Oxford: Oxford University Press, 2006), p. 123.

23 Plato, *Republic* 508b, c; 509b. See Margaret Llasera, 'Concepts of Light in the Poetry of Henry Vaughan', *The Seventeenth Century*, 3 (1988), 47–61 (p. 47).

24 See J. L. Heilbron, 'Introductory Essay', in *John Dee on Astronomy: Propaedeumata Aphoristica (1558 & 1568)* (hereafter *JDA*), ed. and trans. Wayne Shumaker (Berkeley: University of California Press, 1978), pp. 1–99 (p. 36), and Urszula Szulakowska, *The Alchemy of Light: Geometry and Optics in Late Renaissance Alchemical Illustration* (Leiden: Brill, 2000), pp. 33, 40, 48, 57, 65, 68 and 178.

25 See Robert Midgley, *A New Treatise of Natural Philosophy, Free'd from the Intricacies of the Schools* (London, 1687), pp. 161–2, and Kenelm Digby, *Of Bodies, and of Mans Soul to Discover the Immortality of Reasonable Souls*, 2 vols (London, 1669), I, 63.

26 *JDA*, p. 131 (Aphorism XXII). See Nicholas H. Clulee, *John Dee's Natural Philosophy: Between Science and Religion* (London: Routledge, 1988), p. 42.

27 *JDA*, p. 179 (Aphorism XCV), and *A True & Faithful Relation of What Passed for Many Years between Dr. John Dee (A Mathematician of Great Fame in Q. Eliz. and King James their Reignes) and some Spirits* (London, 1659), p. 368.

28 John Dee, 'Mathematicall Preface', appended to *The Elements of Geometrie of the Most Ancient Philosopher Euclide of Megara*, trans. Henry Billingsley (London, 1570), n.p.

29 *JDA*, pp. 171, 181, 183 (Aphorisms XC, XCVI, XCVII, C).

30 See Edward Grant, ed., *Source Book in Medieval Science*, 2 vols (Cambridge, MA: Harvard University Press, 1974), I, 413–14, nn. 104, 105. Dee's own manuscript on burning mirrors (a work he announces in *JDA*, p. 117) partially survives, but is primarily a paraphrase of Antonius Gogova's tract on Alhazen: see Heilbron, 'Introductory Essay', p. 68.

31 See *JDA*, pp. 145, 147, 149, 181 (Aphorisms XLVIII, LII and XCIX).

32 C. H. Josten, 'A Translation of John Dee's *Monas Hieroglyphica* (Antwerp, 1564)', *Ambix*, 12: 2 & 3 (June and October 1964), 84–221 (p. 131).

33 'Testamentum Johannis Dee Philosophu Summi ad Johannem Gwynn, transmissum 1568', in Elias Ashmole, *Theatrum Chemicum Britannicum* (London, 1651), p. 334.

34 *The Alchemist*, 2.2.76, in *The Cambridge Edition of the Works of Ben Jonson* (hereafter *Jonson*), gen. eds. David Bevington and others, 7 vols (Cambridge: Cambridge University Press, 2012), III, 599.

35 *The Alchemist*, 2.6.23-24, in *Jonson*, III, 624, and see Frances A. Yates, *Shakespeare's Last Plays: A New Approach* (London: Routledge & Kegan Paul, 1975), p. 113.

36 *The Alchemist*, 4.1.85-90, in *Jonson*, III, 654–5. See also III, 657 (4.1.139–41).

37 See Kircher, *Oedipus Aegyptiacus*, 3 vols (Rome, 1652–54), II, 21. Kircher also described a form of distillation powered by concentrated rays: see *Mundus Subterraneus*, 2 vols in 1 (Amsterdam, 1665), II, 241.

38 Nicaise Le Févre, *A Compleat Body of Chymistry*, trans. P. D. C., 2 vols (London, 1664), I, 84, 85; II, 241, 242.

39 J. Andrew Mendelsohn, 'Alchemy and Politics in England 1649-1665', *Past & Present*, 135 (May 1992), 30–78 (p. 59); Le Févre, *A Discourse upon Sr Walter Rawleigh's Great Cordial*, trans. Peter Belon (London, 1664), p. 13.

40 See also Le Févre, *A Compleat Body*, II, 134.

41 See Samuel Hartlib, *Ephemerides* (1658), 29/7/1A, in *The Hartlib Papers: A Complete Text and Image Database of the Papers of Samuel Hartlib (c. 1600-1662)*, 2nd edn (HROnline, Humanities Research Institute, University of Sheffield, 2002).

42 Samuel Butler, 'Satire upon the Royal Society', in *Hudibras I and II and Selected Other Writings*, eds. John Wilder and Hugh de Quehen (Oxford: Clarendon Press, 1973), p. 211 (ll. 89–90).

43 John Hall, 'A Burning-glasse', in *Poems* (Cambridge, 1646), p. 19 (ll. 1–6). Hall alludes to Jupiter's retort to Archimedes's planetarium: see Claudian, *Shorter Poems*, 51, in *Greek and Roman Technology*, p. 56. On Hall and Hartlib, see G. H. Turnbull, 'John Hall's Letters to Samuel Hartlib', *Review of English Studies*, n.s. 4: 15 (1953), 221–33.

44 See W. E. Knowles Middleton, 'Archimedes, Kircher, Buffon, and the Burning-Mirrors', *Isis*, 52: 4 (December 1961), 533–43 (p. 534).

45 See Herbert Grabes, *The Mutable Glass: Mirror-Imagery in Titles and Texts of the Middle Ages and English Renaissance* (Cambridge: Cambridge University Press, 1982), pp. 8, 39, and John Garrison, *Glass* (New York: Bloomsbury Academic, 2015), p. 55.

46 Rayna Kalas, *Frame, Glass, Verse: The Technology of Poetic Invention in the English Renaissance* (Ithaca: Cornell University Press, 2007), p. 113.

47 See Vaughan Hart, *Art and Magic in the Court of the Stuarts* (London: Routledge, 1994), p. 155 and *passim*.

48 Charles Edmonds, *Basilikon Doron: Or His Maiesties Instrvctions to His Dearest Sonne, Henry the Prince* (London, 1603), p. 79, sig. A3^r.

49 'Panegyre' (1603), in *Jonson*, II, 475.

50 *Hymenaei* (1605), l. 802, in *Jonson*, II, 698. See also the masque's embodiment of the flame of Reason (ll. 104–9; II, 672). For this emblem, see Thomas Jenner, *The Foure Faculties of the Minde* (London, 1662).

51 See Hart, *Art and Magic*, pp. 181–2, 158–61, 191, 176–8.

52 Roy Strong, *Henry, Prince of Wales and England's Lost Renaissance* (London: Thames and Hudson, 1986), pp. 106–7.

53 Christopher McIntosh, *Gardens of the Gods: Myth, Magic and Meaning in Horticulture* (London: I. B. Tauris, 2005), p. 71. On Memnon's statue, see M. R. Duffey, 'The Vocal Memnon and Solar Thermal Automata', *Leonardo Music Journal*, 17 (2007), 51–4.

54 See Rosalie L. Colie, 'Cornelis Drebbel and Salomon de Caus: Two Jacobean Models for Salomon's House', *Huntington Library Quarterly*, 18: 3 (May 1955), 245–60, William Brenchley Rye, *England as Seen by Foreigners in the Days of Elizabeth and James the First* (London: John Russell Smith, 1865), pp. 232–42, H. A. M. Snelders, 'Drebbel, Cornelis', *Oxford Dictionary of National Biography*, accessed via <https://www.oxforddnb.com>, and Gerritt Tierie, *Cornelius Drebbel* (Amsterdam: H. J. Paris, 1932), p. 50.

55 Drebbel to King James (1612?), trans. Lawrence Ernest Harris, in *The Two Netherlanders: Humphrey Bradley and Cornelis Drebbel* (Leiden: E. J. Brill, 1961), p. 147.

56 See David J. Baker, '"The Allegory of a China Shop": Jonson's "Entertainment at Britain's Burse"', *ELH*, 72: 1 (Spring 2005), 159–80 (p. 161), and Jennifer Speake, 'The Wrong Kind of Wonder: Ben Jonson and Cornelis Drebbel', *Review of English Studies*, 66: 273 (2015), 60–70 (pp. 66, 70).

57 *Britain's Burse*, ll. 53–4, 198–200, 223–4, in *Jonson*, III, 359, 366, 367. Speake suggests that the piece is 'structured by Jonson's attitude' to Drebbel ('The Wrong Kind of Wonder', p. 61). However, the masque not only alludes to the work of de Caus (and John Napier, as suggested below) in addition to Drebbel, but also is concerned with the consumption of wondrous objects generally. On Jonson's knowledge of Drebbel, see Ian Donaldson, *Ben Jonson: A Life* (Oxford: Oxford University Press, 2011), p. 397.

58 *The Masque of Augurs* (1622), ll. 86, 193, 213, 221–4, in *Jonson*, V, 593, 597, 598.

59 See *Jonson*, V, 604n.

60 See Harris, in *The Two Netherlanders*, pp. 189–91.

61 See *Jonson*, III, 364n, and Hart, *Art and Magic*, pp. 181–2.

62 See *Jonson*, III, 364n.

63 See Catherine Rockwood, '"Know Thy Side": Propaganda and Parody in Jonson's *Staple of News*', *ELH*, 75: 1 (Spring 2008), 135–49 (p. 140).

64 See esp. D. F. McKenzie, '*The Staple of News* and the Late Plays', repr. in McKenzie, *Making Meaning: 'Printers of the Mind' and Other Essays*, eds.

Peter D. McDonald and Michael F. Suarez, S.J. (Amherst, MA: University of Massachusetts Press, 2002), pp. 169–97.

65 See Galileo Galilei, 'The Assayer' (1623), in *Discoveries and Opinions of Galileo*, trans. Stillman Drake (New York: Doubleday Anchor Books, 1957), p. 246, and Galileo, *Dialogues Concerning Two New Sciences*, trans. Henry Crew and Alfonso de Salvio (New York: Macmillan, 1914), pp. 41–2.

66 John L. Heilbron, *Galileo* (New York: Oxford University Press, 2010), pp. 234–45.

67 *Jonson*, VI, 82n.

68 See Harris, *The Two Netherlanders*, p. 166.

69 See David McPherson, 'Ben Jonson's Library and Marginalia: An Annotated Catalogue', *Studies in Philology*, 71: 5 (December 1974), 1–106 (p. 43).

70 See Donaldson, *Ben Jonson*, p. 397.

71 John Napier, 'Secrett Inventionis' (7 June 1596), transcribed in Mark Napier, *Memoirs of John Napier of Merchiston, His Lineage, Life, and Times, with a History of the Invention of Logarithms* (Edinburgh: William Blackwood, 1834), p. 247 (and see pp. 254–71).

72 Given Napier's military ambitions for the technology, this identification seems more likely than Speake's suggestion of Drebbel ('The Wrong Kind of Wonder', pp. 66–7).

73 Translation of 'Letter Patent to Mr. William Drummond for the Making of Military Machines' (29 September 1626), in David Masson, *Drummond of Hawthornden: The Story of His Life and Writings* (London: Macmillan, 1873), pp. 157, 159. See Frederick A. Pottle, 'Two Notes on Ben Jonson's *Staple of News*', *Modern Language Notes*, 40: 4 (April 1925), 223–6. In 'Song', Drummond describes the Sun as 'This great and burning Glasse which cleares all Eyes' (*Poems* (London, 1656), p. 61).

74 See Francis F. Madan, *A New Bibliography of the 'Eikon Basilike' of King Charles the First* (Oxford: Oxford University Press, 1950), p. 2.

75 See Christopher Wordsworth, *Documentary Supplement to 'Who Wrote Eikon Basilike?'* (London: J. Murray, 1825), p. 16.

76 See Steven N. Zwicker, *Lines of Authority: Politics and English Literary Culture, 1649-1689* (Ithaca, NY: Cornell University Press, 1993), p. 41.

77 See Elizabeth Skerpan Wheeler, '*Eikon Basilike* and the Rhetoric of Self-Representation', in *The Royal Image: Representations of Charles I*, ed. Thomas N. Corns (Cambridge: Cambridge University Press, 1999), pp. 122–40 (p. 134).

78 See, for instance, Michael Maier, *Septimana Philosophica* (Frankfurt, 1620), p. 31.

79 Ernest B. Gilman, *Iconoclasm and Poetry in the English Reformation* (Chicago: The University of Chicago Press, 1986), p. 154.

80 Kevin Sharpe, 'The Royal Image: An Afterword', in *The Royal Image: Representations of Charles I*, ed. Thomas N. Corns (Cambridge: Cambridge University Press, 1999), pp. 289–309 (p. 291). On the *text*'s use of optics, vision and perspective, see Laura L. Knoppers, 'Imagining the Death of the King: Milton, Charles I, and Anamorphic Art', in *Imagining Death in Spenser and Milton*, eds. Elizabeth Jane Bellamy, Patrick Cheney and Michael Schoenfeldt (Houndmills: Palgrave Macmillan, 2003), pp. 151–70 (esp. pp. 158–61).

81 *Eikon Basilike* (London, 1648), p. 70. See also p. 200. The book also compares the king's executed adviser Thomas Wentworth (the Earl of Strafford) to the Sun, since he had 'so vigorous a lustre' (p. 6): a simile which provoked Milton's derision, since the Sun 'beares allusion to a King, not to a Subject' (*Eikonoklastes*, 2nd edn (London, 1650), p. 20).

82 See, for instance, Henry Vaughan, 'A King Disguis'd', in *Thalia Rediviva: The Pass-Times and Diversions of a Countrey-Muse* (London, 1678), p. 2.

83 William Lilly, *The Starry Messenger* (London, 1645), pp. 19, 10, 11. See also Lilly, *Monarchy or No Monarchy in England* (London, 1651), p. 118, and Ann Geneva, *Astrology and the Seventeenth-Century Mind: William Lilly and the Language of the Stars* (Manchester: Manchester University Press, 1995), p. 268.

84 Thomas Povey, *The Moderator Expecting Sudden Peace, or Certaine Ruine* (London, 1642), p. 14.

85 See John Gauden, *Considerations Touching the Liturgy of the Church of England*, 2nd edn (London, 1661), p. 15.

86 See Jeremiah Burroughs, *The Saints Treasury Being Sundry Sermons Preached in London* (London, 1654), p. 48.

87 *The Life of S. Augustine: The First Part* (London, 1660), p. 155 (Book IX, Chap. IV); Anthony Burgess, *Spiritual Refining: Or A Treatise of Grace and Assurance* (London, 1652), p. 30 (see also p. 171); John Flavel, *The Fountain of Life Opened, Or, A Display of Christ in His Essential and Mediatorial Glory* (London, 1673), p. 338.

88 Deborah Madden, 'Medicine and Moral Reform: The Place of Practical Piety in John Wesley's *Art of Physic*', *Church History*, 73: 4 (December 2004), 741–58 (p. 748).

89 Thomas Birch, ed., *The History of the Royal Society of London for Improving of Natural Knowledge*, 4 vols (London, 1756–7), III, 4 (18 January 1671/2).

90 Anthony Horneck, *The Great Law of Consideration: Or a Discourse*, 2nd edn (London, 1678), p. 13.

91 See Charles Clay Doyle, 'Seeing through Colored Glasses', *Western Folklore*, 60: 1 (Winter 2001), 67–91 (p. 79).

92 See William W. E. Slights, *The Heart in the Age of Shakespeare* (Cambridge: Cambridge University Press, 2008), pp. 6–7, 104.

93 See Scott Manning Stevens, 'Sacred Heart and Secular Brain', in *The Body in Parts: Fantasies of Corporeality in Early Modern Europe*, eds. David Hillman and Carla Mazzio (New York: Routledge, 1997), pp. 262–82 (p. 273).

94 'Against the Heavenly Prophets in the Matter of Images and Sacraments (1525), Part I', trans. Bernhard Erlling, in *Luther's Works*, ed. Conrad Bergendorff, American edn (Philadelphia: Muhlenberg Press, 1958), 40: 99–100.

95 W. Reginald Ward and Richard P. Heitzenrater, eds., *The Works of John Wesley*, 23 vols (Nashville: Abingdon Press, 1988), XVIII: Journals and Diaries I (1735–8), 250.

96 William Harvey, *Movement of the Heart and Blood in Animals*, trans. Kenneth J. Franklin (Oxford: Blackwell, 1957), p. 59.

97 *The Life of St Teresa of Ávila by Herself*, trans. J. M. Cohen (London: Penguin, 1957), pp. 210, 146.

98 See Charles Partee, *The Theology of John Calvin* (Louisville, KY: Westminster John Knox Press, 2008), p. xiv.

99 Cf. the '*Smoking Heart*' illustration in George Wither, *A Collection of Emblemes, Ancient and Moderne* (London, 1635), p. 39.

100 John Suckling, 'Love's Burning-glass', ll. 3-6, 9-14, in *The Works of Sir John Suckling*, ed. Thomas Clayton, 2 vols (Oxford: Clarendon Press, 1971), I, 32.

101 Suckling, 'Loves World', ll. 1–4, 13–16, 29–30, in *Works*, I, 22. See Robert Wilcher, *The Discontented Cavalier: The Work of Sir John Suckling in Its Social, Religious, Political, and Literary Contexts* (Newark: University of Delaware Press, 2007), p. 97.

102 Cf. William Fairfax, 'The Union', ll. 9–10, trans. Thomas Stanley (1651), in *The Poems and Translations of Thomas Stanley*, ed. Galbraith Miller Crump (Oxford: Clarendon Press, 1962), p. 66.

103 Katherine Philips, 'To the Excellent Mrs. Anne Owen, Upon Her Receiving the Name of Lucasia, and Adoption into Our Society, December 28, 1651', ll. 7–14, in *Minor Poets of the Caroline Period*, ed. George Saintsbury, 3 vols (Oxford: Clarendon Press, 1905–21), I, 526.

104 John Hopkins, 'To Amasia, holding a Burning-Glass in her Hand', ll. 1–8, in *Amasia, or, The Works of the Muses* (1700), 2 vols (London, 1700), I, Book II, p. 51.

105 See John Cleveland, 'To the State of Love, Or the Senses' Festival' (1651), ll. 22–3, in *Minor Poets of the Caroline Period*, III, 20, and William Congreve,

Incognita: Or, Love and Duty Reconcil'd. A Novel (London, 1692), p. 113. On the French tradition, see Elise Goodman-Soellner, 'Nicolas Lancret's *Le miroir ardent*: An Emblematic Image of Love', *Simiolus: Netherlands Quarterly for the History of Art*, 13: 3/4 (1983), 218–24 (p. 220).

106   Thomas D'Urfey, *The Fool Turn'd Critick: A Comedy* (London, 1678), p. 23 (2.3.127–28). See also D'Urfey, *The Comical History of Don Quixote: The Third Part* (London, 1696), p. 41 (4.2).

107   John Cleland, *Memoirs of a Woman of Pleasure, or; Fanny Hill*, ed. Peter Wagner (London: Penguin, 1985), p. 219.

108   See Anna Marie Roos, *Luminaries in the Natural World: Perceptions of the Sun and Moon in England, 1400-1720* (New York: Peter Lang, 2001), pp. 2, 3, 28.

109   See Marjorie Nicolson, *Voyages to the Moon* (New York: Macmillan, 1948), pp. 160–7.

110   Cyrano de Bergerac, *The Comical History of the States and Empires of the Moon and Sun*, trans. Archibald Lovell (London, 1687), pp. 4, 5. An earlier, fragmentary, English translation depicts the first aerial invention on its frontispiece: *Selenarhia, or, The Government of the World in the Moon: A Comical History*, trans. Thomas St. Serf (London, 1659).

111   Frédérique Aït-Touati suggests Cyrano's novels seem 'unconcerned about defending the Copernican system' (*Fictions of the Cosmos: Science and Literature in the Seventeenth Century*, trans. Susan Emanuel (Chicago: University of Chicago Press, 2011), p. 71). However, the foregrounding of solar devices and deification of the Sun in *The Comical History* constitutes an implicit assertion of heliocentrism.

112   *The Elements of Geometrie*, p. 339.

113   On the Sun as Cyrano's 'visible God', see Erica Harth, *Cyrano de Bergerac and the Polemics of Modernity* (New York: Columbia University Press, 1970), esp. p. 140.

Chapter 2

1   Aphra Behn, *Oroonoko*, in *The Works of Aphra Behn*, ed. Janet Todd, 7 vols (London: Pickering & Chatto, 1995), III, 59.

2   See esp. Albert J. Rivero, 'Aphra Behn's *Oroonoko* and the "Blank Spaces" of Colonial Fictions', *Studies in English Literature, 1500-1900*, 39 (1999), 443–62 (p. 456).

3   It was unknown at this time that the Olmecs of Mexico had used burning-mirrors possibly as early as 1200 BC: see Goldberg, *The Mirror and Man*, p. 80.

4   See Peter Fidler, 'Journal of a Journey Overland from Buckingham House to the Rocky Mountains in 1792 and 3', MSS. E.3/2, Hudson's Bay Company Archives, cited in John C. Ewers, 'When Red and White Men Met', *The Western Historical Quarterly*, 2: 2 (April 1971), 133–50 (p. 137).

5   See John Wesley, *A Survey of the Wisdom of God in the Creation*, 2 vols (Bristol, 1763), I, 13.

6   See Jean Haudicquer de Blancourt, *The Art of Glass* (London, 1699), pp. 345–51.

7   William Derham, *Astro-Theology: or, a Demonstration of the Being and Attributes of God, from a Survey of the Heavens* (London, 1715), p. 152. See also Derham, *Physico-Theology; or, a Demonstration of the Being and Attributes of God, from His Works of Creation* (London, 1713), p. 46n, and John Laurence, *A New System of Agriculture* (London, 1726), p. 18.

8   Henry Justel to Oldenburg, [?] August 1665, in *The Correspondence of Henry Oldenburg* (hereafter *Corr. Oldenburg*), eds. A. Rupert Hall and Marie Boas Hall, 13 vols (Madison: University of Wisconsin Press, 1965–86), II, 456.

9   'An Account of a Not ordinary *Burning Concave*, Lately Made at *Lyons*', *Philosophical Transactions* (hereafter *PT*), 1 (1665), 95–8 (pp. 96, 97).

10 See *Corr. Oldenburg*, II, 458, 475, 477, 485, 544–6; V, 62, 64, and 'An Extract of a Letter from *Paris*, about … the Making of an Extraordinary *Burning-glass* at *Milan*', *PT*, 3 (1668), 795–6 (p. 796).

11 'An Account of the Invention of Grinding *Optick* and *Burning*-Glasses, of a Figure Not-*Spherical*', *PT*, 3 (1667/8), 631–2; *Corr. Oldenburg*, IV, 223; *The Diary of Samuel Pepys*, eds. Robert Latham and William Matthews, 11 vols (London: Bell & Hyman, 1970–83), IX, 113 (12 March 1667/8).

12 See *Corr. Oldenburg*, VI, 162, 163–4, and also P. Francesco Lana, 'Some of the effects of the *Burning Concave* of *Lions*', *PT*, 6 (1671/2), 3060.

13 'An Account from *Paris* Concerning a great Metallin *Burning Concave*, and Some of the most considerable Effects of it', *PT*, 4 (1669), 986–7 (p. 986).

14 Indra Kagis McEwen, 'Midsummer Moderns: The Foundation of the Paris Observatory, 21 June 1667', in *Foundation, Dedication and Consecration in Early Modern Europe*, eds. Maarten Delbeke and Minou Schraven (Leiden: Brill, 2012), pp. 335–62 (pp. 356–9).

15 Sébastien Leclerc, frontispiece to Claude Perrault, *Mémoires pour Servir à l'Histoire Naturelle des Animaux*, 2 vols (Paris, 1671–6), I.

16 Sabine Melchior-Bonnet, *The Mirror: A History*, trans. Katharine H. Jewett (New York: Routledge, 2001), p. 178.

17 Here I borrow terms from Jeffrey C. Alexander, 'Iconic Experience in Art and Life: Surface/Depth Beginning with Giacometti's *Standing Woman*', *Theory, Culture & Society*, 25 (2008), 1–19 (pp. 6–7).

18 Francis Vernon to Oldenburg, 20 October 1669, in *Corr. Oldenburg*, VI, 295. See della Porta, *Natural Magick*, p. 361.

19 Anon., 'Robertson, Artist in Ghosts', *Household Words*, 10: 253 (27 January 1855), 553–8 (p. 553), translating Étienne-Gaspard Robertson, *Mémoires Récréatifs Scientifiques et Anecdotiques*, 2 vols (Paris: by the author and Librairie de Wurtz, 1831), I, 115.

20 See Emily Jane Cohen, 'Enlightenment and the Dirty Philosopher', *Configurations*, 5: 3 (1997), 369–424 (esp. pp. 382–3).

21 In June 1681, for instance, the mathematician Robert Wood (1621?–85) presented a paper on the 'phænomena of a burning glass'. See Birch, *The History of the Royal Society of London for Improving of Natural Knowledge*, IV, 35, 71, 93, 114. Wood's paper appears to survive as the hitherto unidentified 'Observations on ye Burninglass' [n.d.], Royal Society MS, Classified Papers 1660-1740, 31 vols, II, f. 12.

22 'A Relation of the Great Effects of a New Sort of Burning *Speculum* Lately Made in *Germany*', *PT*, 16: 188 (1687), 352–4 (pp. 353–4).

23 Fokko Jan Dijksterhuis, 'Foci of Interests. Optical Pursuits amongst Huygens, Leibniz and Tschirnhaus 1680-1710', in *Der Philosoph im U-Boot: praktische Wissenschaft und Technik im Kontext von Gottfried Wilhem Leibniz*, ed. Michaal Kempe (Hannover: Gottfried Wilhelm Leibniz Bibliothek, 2015), pp. 261–83 (pp. 264, 262). On Settala's mirror, see 'An Extract Of a Letter from *Paris*, about the polishing of *Telescopical* Glasses by a *Turn-lathe*; as also the making of an extraordinary *Burning-glass* at *Milan*', *PT*, 3: 40 (19 October 1668), 795–6 (p. 796).

24 Bernard Le Bovier de Fontenelle, *The Lives of the French, Italian and German Philosophers, Late Members of the Royal Academy of Sciences in Paris*, trans. anon (London, 1717), p. 135.

25 Fontenelle, *Lives*, p. 141. See also Dijksterhuis, 'Foci of Interests', pp. 274–5.

26 Ehrenfried Walter von Tschirnhaus, 'Singularia effecta vitri caustici bipedalis', *Acta eruditorum* (November 1691), 517–20; Fontenelle, *Lives*, p. 142.

27 Edmund de Waal, *The White Road* (London: Chatto & Windus, 2015), p. 190.

28 Johann Martin Bernigeroth, portrait of Tschirnhaus (1708), Staatlichte Kunstgammlungen Dresden, Kupferstich-Kabinett, in Peter Plassmeyer and others, *Ehrenfried Walther von Tschirnhaus (1651-1708): Experiment emit dem Sonnenfeuer* (Dresden: Staatliche Kunstammlungen Dresden, 2001), p. 118.

29 See Etienne François Geoffroy to Hans Sloane, 10 August 1703, BL Sloane MSS 4039, ff. 171–2.

30 David Gregory, *Isaac Newton and Their Circle: Extracts from David Gregory's Memoranda, 1677-1708*, ed. W. G. Hiscock (Oxford: Oxford University Press, 1937), p. 17.

31 Etienne François Geoffroy, 'Experiments upon Metals, made with the Burning-Glass of the Duke of *Orleans*', *PT*, 26 (1709), 374–86 (pp. 374, 376, 378). See Frederic L. Holmes, 'The Communal Context for Etienne François Geoffroy's "Table des rapports"', *Science in Context*, 9 (1996), 289–311 (esp. pp. 294, 306).

32 Royal Society MS, Journal Book (Copy), 43 vols, X, 58 (20 January 1704).

33 John Evelyn, *Diary*, ed. E. S. de Beer, 6 vols (Oxford: Clarendon Press, 1955), V, 592 (22 April 1705). See also *Diary*, V, 600 (17 June 1705).

34 See Royal Society MS, Journal Book (Copy), X, 75–6 (17 and 24 May 1704).

35 See esp. Simon Schaffer, 'Glass Works', in *The Uses of Experiment: Studies in the Natural Sciences*, eds. D. Gooding, Trevor Pinch and Simon Schaffer (Cambridge: Cambridge University Press, 1989), pp. 67–104.

36 See Evelyn, *Diary*, V, 592 (22 April 1705), and Derham, *Astro-Theology*, p. 153n.

37 Gregory, 19 May 1704 and 23 June 1705, in *Isaac Newton and Their Circle*, pp. 17, 26; John Harris, *Lexicon Technicum: Or, an Universal English Dictionary of Arts and Sciences*, 2nd edn, 2 vols (London, 1708), II, s.v. 'BURNING Glasses'.

38 A. D. C. Simpson, 'Newton's Telescope and the Cataloguing of the Royal Society's Repository', *Notes and Records of the Royal Society of London*, 38 (1984), 187–214 (pp. 200–1).

39 See D. L. Simms and P. L. Hinkley, 'Brighter than How Many Suns? Sir Isaac Newton's Burning Mirror', *Notes and Records of the Royal Society of London*, 43 (1989), 31–51 (pp. 42–3), and also Simms and Hinkley, 'David Gregory on Newton's Burning Mirror', *Notes and Records of the Royal Society of London*, 55 (2001), 185–90.

40 C. A. Van Peursen, 'E. W. Von Tschirnhaus and the Ars Inveniendi', *Journal of the History of Ideas*, 54: 3 (July 1993), 395–410 (p. 399).

41 'A Letter of Mr. Isaac Newton, ... containing his New Theory about Light and Colors', *PT*, 6 (1671/2), 3075–87 (p. 3079).

42 See G. N. Cantor, *Optics after Newton: Theories of Light in Britain and Ireland, 1704-1840* (Manchester: Manchester University Press, 1983), p. 30, and Isaac Newton *Opticks*, 4th edn (London, 1730), p. 345 (Book III, Query 29).

43 William Molyneux, *Dioptrica Nova* (London, 1692), pp. 198, 200; Harris, *Lexicon Technicum*, I, s.v. 'Light'.

44 See Wilhelm Homberg, 'Observations faites par le moyen du verre ardent', *Mémoires de l'Académie Royale des Sciences*, 4 (1702), 141–9, and *Memoirs of the Royal Academy of Science in Paris Epitomized*, 2nd edn (London, 1721), p. 286.

45 See Dijksterhuis, 'Foci of Interests', p. 279.

46 See Homberg, *Memoirs of the Royal Academy of Science*, pp. 286 and 327.

47 Reported in Edward King, *Morsels of Criticism* (London, 1788), p. 65.

48 I adapt Roland Barthes's notion: see 'The Reality Effect' (1968), in *The Rustle of Language*, trans. Richard Howard (Berkeley: University of California Press, 1989), pp. 141–8.

49 On the wider issues, see G. N. Cantor, 'Weighing Light: The Role of Metaphor in Eighteenth-Century Optical Discourse', in *The Figural and the Literal: Problems of Language in the History of Science and Philosophy, 1630-1800*, eds. Andrew E. Benjamin, G. N. Cantor and John R. R. Christie (Manchester: Manchester University Press, 1987), pp. 124–46.

50 See, for example, Robert Boyle, *Occasional Reflections upon Several Subjects* (1665), pp. 216–17, in *The Works of Robert Boyle*, eds. Michael Hunter and Edward B. Davis, 14 vols (London: Pickering & Chatto, 1999), V, 178, and Edward Young, *The Works of the Author of the Night-Thoughts*, 4 vols (London, 1757), IV, 239.

51 Joseph Addison, *The Tatler*, 100 (29 November 1709), in *The Tatler*, ed. Donald F. Bond, 3 vols (Oxford: Clarendon Press, 1987), II, 114, 115.

52 See Thomas Creech, *T. Lucretius Carus The Epicurean Philosopher, His Six Books De Natura Rerum Done into English Verse* (London, 1682), p. 4.

53 This reverses the conceit in Aristophanes's *The Clouds* (423BC), in which the protagonist Strepsiades tries to use a burning-glass to scorch the record of his debts from the court's docket. See *The Clouds*, ed. K. J. Dover (Oxford: Clarendon Press, 1968), pp. 47, 194n.

54 See Grabes, *The Mutable Glass*, p. 99, and Jonathan Swift, *A Tale of a Tub and Other Works*, ed. Marcus Walsh (Cambridge: Cambridge University Press, 2010), p. 142.

55 Francesco Cepparuli, 'Truth Opens the Eyes of the Blind' (1744), frontispiece to Giussepe Antonio Costantini, *Lettere critiche, giocose, morali, scientifiche, ed erudite alla moda, ed al guste del Secolo presente*, 7 vols (Venice, 1751), Vol. I. See Rolf Reichardt and Deborah Louise Cohen, 'Light against Darkness: The Visual Representations of a Central Enlightenment Concept', *Representations*, 61 (Winter 1998), 95–148 (pp. 105–6).

56 For the poetry, see Chapter 1. For visual depictions, see Cæsar Ripa, *Iconologia: Or, Moral Emblems* (London, 1709), pp. 54–5; Nicolas Lancret, *Le miroir ardent* (ca. 1730), Berlin, Schloss Charlottenburg; Charles-Joseph Natoire, *La Beauté Rallume le Flambeau de l'amour* (1739), Versailles, Musée national du château et des Trianons, MV8349.

57 Pope to Arbuthnot, 11 July (1714), in *The Correspondence of Dr. John Arbuthnot* (hereafter *Corr. Arbuthnot*), ed. Angus Ross (München: Fink, 2006), pp. 186–7.

58 See Marjorie Hope Nicolson and G. S. Rousseau, *'This Long Disease, My Life':*
    Alexander Pope and the Sciences (Princeton: Princeton University Press, 1968),
    p. 268, and compare Newton, *Opticks*, p. 17.

59 Swift to Arbuthnot, 25 July 1714, in *Corr. Arbuthnot*, p. 195.

60 George Lakoff and Mark Johnson, *Metaphors We Live By* (Chicago: University
    of Chicago Press, 1980), p. 97.

61 Arbuthnot to Swift, 17 July 1714, in *Corr. Arbuthnot*, p. 191.

62 *A Letter to a Young Poet* (1721), in *The Prose Works of Jonathan Swift*, eds.
    Herbert Davis and others, 14 vols (Oxford: Blackwell, 1939–68), IX, 334;
    Charles Kerby-Miller, ed., *The Memoirs of the Extraordinary Life, Works, and*
    Discoveries of Martinus Scriblerus (New Haven, CT: Yale University Press,
    1950), p. 167.

63 Swift to Archdeacon Walls, 30 March 1717, in *The Correspondence of Jonathan*
    Swift, D. D., ed. David Woolley, 4 vols (Frankfurt am Main: Lang, 1999–2007),
    II, 237.

64 See Swift to Arbuthnot, 3 July 1714 and 25 July 1714, in *Corr. Arbuthnot*,
    pp. 182, 195.

65 See John Conduitt, 'Miscellanea', in King's College, Cambridge, Keynes MS
    130.5, f. 4^v.

66 George A. Aitken, ed., *The Life and Works of John Arbuthnot* (Oxford:
    Clarendon Press, 1892), p. 426. The BL copy of the *Essay* (618.c.28) is ascribed
    to Arbuthnot by a contemporary hand.

67 John Arbuthnot, *To the Right Honourable the Mayor and Aldermen of the City of*
    London, p. 1.

68 See Edmond Halley, 'Observations of the Late Total Eclipse of the Sun on the
    22nd of April 1st Past', *PT*, 29 (1715), 245–62. Such satires include *A True and*
    Faithful Narrative of What Passed in London (1732): see *Poetry and Prose of John*
    Gay, ed. Vinton A. Dearing, 2 vols (Oxford: Clarendon Press, 1974), II, 473.

69 See Christine MacLeod, *Inventing the Industrial Revolution: The English Patent*
    System, 1660-1800 (Cambridge: Cambridge University Press, 1988), p. 2.

70 See Richard Mead, *Of the Power and Influence of the Sun and Moon on Humane*
    Bodies (London, 1712), pp. 43–4, and Anna Marie Roos, 'Luminaries in
    Medicine: Richard Mead, James Gibbs, and Solar and Lunar Effects on the
    Human Body in Early Modern England', *Bulletin of the History of Medicine*, 74
    (2000), 433–57 (pp. 437–8).

71 George Cheyne, *Philosophical Principles of Natural Religion* (London, 1705),
    p. 96 (see also p. 98). Arbuthnot recommended Cheyne's book to Newton: see
    John Conduitt, 'Character', King's College, Cambridge, Keynes MS 130.7, f. 2^r.

72   See esp. G. S. Rousseau, 'Wicked Whiston and the English Wits', in
     Enlightenment Borders: Pre- and Post-Modern Discourses: Medical, Scientific
     (Manchester: Manchester University Press, 1991), pp. 325–41 (p. 333).

73   See esp. Larry Stewart, *The Rise of Public Science: Rhetoric, Technology, and
     Natural Philosophy in Newtonian Britain, 1660-1750* (Cambridge: Cambridge
     University Press, 1992).

74   *A Description of the Great Burning-Glass Made by Mr. Villette and His Two Sons,
     Born at Lyons. With Some Remarks upon the Surprising and Wonderful Effects
     Thereof* (London, 1718), p. 7. The text adapts and translates *Description du
     Grand Miroir Ardent, fait par les Sieurs Villette Pere et Fils Natifs de Lion* (Liège,
     1715).

75   Jonathan Swift, *Gulliver's Travels*, ed. David Womersley (Cambridge: Cambridge
     University Press, 2012), pp. 259–60 (Book III, Chapter V).

76   Swift knew personally at least one projector in desperate financial and
     psychological circumstances. See Gregory Lynall, 'Scriblerian Projections of
     Longitude: Arbuthnot, Swift, and the Agency of Satire in a Culture of Invention',
     Journal of Literature and Science, 7: 2 (2014), 1–18 (p. 11).

77   On the similarity of this experimenter's spiel to genuine 'project' literature, see
     David Alff, 'Swift's Solar Gourds and the Rhetoric of Projection', *Eighteenth-
     Century Studies*, 47: 3 (Spring 2014), 245–60 (esp. p. 252).

78   See Marjorie Nicolson and Nora M. Mohler, 'The Scientific Background of
     Swift's *Voyage to Laputa*', *Annals of Science*, 2 (1937), 299–334 (p. 301).

79   Robertson, *Mémoires*, I, 118.

80   *A Description of the Great Burning-Glass*, pp. 3–4.

81   *Daily Courant*, 16 July 1719.

82   John Harris and John Theophilis Desaguliers, 'An Account of Some
     Experiments Tried with Mons. Villette's Burning Concave, in June 1718', *PT*, 30
     (1719), 976–7.

83   *A Description of the Great Burning-Glass*, pp. 12–13.

84   Jeffrey R. Wigelsworth, *Selling Science in the Age of Newton: Advertising and the
     Commoditization of Knowledge* (Farnham: Ashgate, 2010), p. 66.

85   *Post Man and the Historical Account*, 1878 (3–5 November 1720).

86   See Johnson to Mrs Thrale, 30 June 1783, in *The Letters of Samuel Johnson*, ed.
     Bruce Redford, 5 vols (Oxford: Clarendon Press, 1992–4), IV, 161.

87   See, for instance, James Parsons, 'Observations upon Father Kircher's Opinion
     Concerning the Burning of the Fleet of Marcellus by Archimedes', *PT*, 48
     (1753–4), 621–5, and Patrick Brydone, *A Tour through Sicily and Malta*, 2 vols
     bound as 1 (London, 1773), pp. 272–3.

88 Turberville Needham, 'Of a New Mirror, Which Burns at 66 Feet Distance, Invented by M. De Buffon', *PT*, 44 (1747), 493–5 (p. 494).

89 *Barr's Buffon. Buffon's Natural History*, trans. J. S. Barr, 10 vols (London, 1797), X, 193–244; Marquis Nicolini, 'Concerning the Same Mirror Burning at 150 Feet Distance', *PT*, 44 (1747), 495–6 (p. 495).

90 Needham, 'Of a new Mirror', p. 493. See also Edward Gibbon, *The History of the Decline and Fall of the Roman Empire*, 12 vols (London, 1788), IV, 90 (Chap. XL, Part V).

91 See Thomas L. Hankins and Robert J. Silverman, *Instruments and the Imagination* (Princeton, NJ: Princeton University Press, 1995).

92 Edmund Burke, *A Philosophical Enquiry into the Origin of Our Ideas of the Sublime and Beautiful*, 2nd edn (London, 1759), pp. 145, 146.

93 Pierre Joseph Macquer, *Elements of the Theory and Practice of Chymistry*, trans., 5th edn (Edinburgh, 1777), p. 7. See also Emilie Du Châtelet, 'Dissertation on the Nature and Propagation of Fire' (1738), in *Selected Philosophical and Scientific Writings*, ed. Judith P. Zinsser, trans. Isabelle Bour and Judith P. Zinsser (Chicago: University of Chicago Press, 2009), p. 72.

94 See Birch, *The History of the Royal Society*, IV, 470 (31 March 1686), Francis Hauksbee, 'An Account of an Experiment, Touching the Quantity of Air Produced from a Certain Quantity of Gunpowder Fired in Common Air', *PT*, 25: 311 (1707), 2409–11, and Hooke, *Micrographia* (London, 1665), p. 174.

95 See Henry Guerlac, *Lavoisier—The Crucial Year: The Background and Origin of His First Experiments on Combustion in 1772* (New York: Gordon and Breach, 1990), pp. 157–8.

96 W. A. Smeaton, 'Pierre Joseph Macquer: Early Attempts to Melt Platinum', *Platinum Metals Review*, 28: 1 (1984), 25–30.

97 See *Reflexions sur les experiences qu'on peut tenter a l'aide du miroir ardent*, or *August Memorandum* (8 August 1772), in *Oeuvres de Lavoisier*, ed. Édouard Grimaux, 6 vols (Paris: Imprimerie Imperiale, 1862–93), III, 261–348, Robert Siegfried, 'Lavoisier's View of the Gaseous State and Its Early Application to Pneumatic Chemistry', *Isis*, 63: 1 (March 1972), 59–78 (p. 69), and C. E. Perrin, 'Document, Text and Myth: Lavoisier's Crucial Year Revisited', *British Journal for the History of Science*, 22 (1989), 3–25 (pp. 8–10, 20).

98 See Priestley, *The History and Present State of Discoveries relating to Vision, Light, and Colours* (London, 1772), esp. pp. 120–2, 169–71, 228–30, 453, 749.

99 See Lawrence Badash, 'Joseph Priestley's Apparatus for Pneumatic Chemistry', *Journal for the History of Medicine and Allied Sciences*, 19 (1964), 139–55 (p. 149).

100 Joseph Priestley, *Experiments and Observations on Different Kinds of Air,*
     and Other Branches of Natural Philosophy, Connected with the Subject, 3 vols
     (Birmingham, 1790), II, 431.

101 W. H. Brock, 'Joseph Priestley: Enlightened Experimentalist', in *Joseph Priestley:*
     Scientist, Philosopher, and Theologian, eds. Isabel Rivers and David L. Wykes
     (Oxford: Oxford University Press, 2008), pp. 49–79 (pp. 59–60).

102 See esp. Robert E. Schofield, *The Enlightened Joseph Priestley: A Study of His Life*
     and Work from 1773 to 1804 (University Park, PA: Pennsylvania State University
     Press, 2004), pp. 105–19.

103 Priestley to Josiah Wedgwood, May 1785, in *Scientific Correspondence of Joseph*
     Priestley, ed. Henry Carrington Bolton (New York: privately printed, 1892),
     p. 76. See also *Memoirs of Dr. Joseph Priestley* (London: for J. Johnson, 1806),
     pp. 93–4.

104 See Joseph Priestley, *Experiments and Observations on Different Kinds of Air*, 2
     vols (London, 1775), II, xxxv–xxxvii, and Badash, 'Joseph Priestley's Apparatus',
     pp. 144–5.

105 See Priestley to Wedgwood, 6 March and 16 September 1782, in *Scientific*
     Correspondence, pp. 33, 39, and Schofield, *The Enlightened Joseph Priestley*, p.
     114.

106 See James Bryant Conant, 'The Overthrow of the Phlogiston Theory: The
     Chemical Revolution of 1775-1789', in *Harvard Case Histories in Experimental*
     Science, 2 vols (Cambridge, MA: Harvard University Press, 1957; repr. 1964),
     I, 68–115, and C. E. Perrin, 'Research Traditions, Lavoisier, and the Chemical
     Revolution', *Osiris*, 4 (1988), 53–81.

107 Badash, 'Joseph Priestley's Apparatus', p. 145.

108 Ursula Klein, 'Technoscience avant la lettre', *Perspectives on Science*, 13 (2005),
     226–66 (p. 227).

109 Abraham Rees, ed., *The Cyclopaedia, or, Universal Dictionary of Arts, Sciences*
     and Literature, 45 vols (London: Longman, 1819–20), V, s.v. 'Burning-*glass*, or
     burning-mirror', and W. A. Smeaton, 'Some Large Burning Lenses and Their
     Use by Eighteenth-Century French and British Chemists', *Annals of Science*, 44
     (1987), 265–76 (pp. 271–3). See also James Millar, ed., *Encyclopaedia Britannica;*
     or, A Dictionary of Arts and Sciences, 4th edn, 20 vols (Edinburgh: Andrew Bell,
     1810), III, 787–8, and David Brewster, ed., *The Edinburgh Encyclopaedia*, 18 vols
     (Edinburgh: for William Blackwood and others, 1830), V, 141–3.

110 William Parker to Banks, 4 and 16 July 1782, in *The Scientific Correspondence of*
     Sir Joseph Banks, ed. Neil Chambers, 6 vols (London: Pickering & Chatto, 2007),
     I, 337, 341.

111 Rees, *Cyclopaedia*; William Alexander, 'Journal of a Voyage to Pekin in China', BL Add MS 35174, f. 86.

112 See J. L. Cranmer-Byng and Trevor H. Levere, 'A Case Study in Cultural Collision: Scientific Apparatus in the Macartney Embassy to China, 1793', *Annals of Science*, 38 (1981), 503–25 (p. 505), and Bernard Lightman, Gordon McOuat and Larry Stewart, 'Introduction', in *The Circulation of Knowledge Between Britain, India and China*, eds. B. Lightman, G. McOuat and L. Stewart (Leiden: Brill, 2013), pp. 1–17 (pp. 2, 10).

113 George Macartney, *An Embassy to China; Being the Journal Kept by Lord Macartney during His Embassy to the Emperor Ch'ien-lung, 1793-1794*, ed. J. L. Cranmer-Byng (London: Longman, 1962), pp. 69, 145–6 (25 July 1793, 1 October 1793). It is unclear whether Hongli's court had encountered thermal optics previously, although the ancient Chinese had used burning-mirrors from possibly as early as 1000 BC.

114 William Proudfoot, *Biographical Memoir of James Dinwiddie, L.L.D.* (Liverpool: Edward Howell, 1868), pp. 53, 54.

115 See Martin Fitzpatrick, 'Priestley in Caricature', in *Oxygen and the Conversion of Future Feedstocks* (London: Royal Society of Chemistry, 1984), pp. 345–69 (esp. pp. 347, 349).

116 *The Priestley Memorial at Birmingham, August 1874* (London: Longman, 1875), prefatory note (n.p.), and p. 147.

117 *The Priestley Memorial*, pp. 1, 32. Dawson perhaps had in mind Priestley's description of mental concentration, in which a burning-glass metaphor is implicit: see Chapter 3.

118 Christine MacLeod, *Heroes of Invention: Technology, Liberalism and British Identity, 1750-1914* (Cambridge: Cambridge University Press, 2007), pp. 24, 307, 291, 310.

119 George Noszlopy, *Public Sculpture of Birmingham: Including Sutton Coldfield*, ed. Jeremy Beach (Liverpool: Liverpool University Press, 1998), p. 34.

120 On the wider rhetorical connections, see Geoffrey Cantor, 'Light and Enlightenment: An Exploration of Mid-Eighteenth-Century Modes of Discourse', in *The Discourse of Light from the Middle Ages to the Enlightenment: Papers Read at a Clark Library Seminar, 24 April 1982*, eds. David C. Lindberg and Geoffrey Cantor (Los Angeles, CA: William Andrews Clark Memorial Library, University of California, 1985), pp. 67–106.

121 Priestley, *Experiments and Observations on Different Kinds of Air*, second edn corrected, pp. xiii–xiv. See Simon Schaffer, 'Priestley's Questions: An Historiographic Survey', *History of Science*, 22 (June 1984), 151–83 (p. 173).

122 Priestley, *Experiments and Observations Relating to Various Branches of Natural Philosophy*, 2 vols (Birmingham, 1781), II, ix.

123 'Eighteen Hundred and Eleven', in *The Works of Anna Laetitia Barbauld*, 2 vols (London, 1825), I, 247 (ll. 261–4). See also Barbauld, 'Address to the Opposers of the Repeal of the Corporation and Test Acts', in *The Works of Anna Laetitia Barbauld*, II, 371.

124 Hannah More, *Slavery: A Poem* (London, 1788), pp. 1, 2 (ll. 2, 16, 17–18).

125 See esp. Terry Castle, 'Phantasmagoria: Spectral Technology and the Metaphorics of Modern Reverie', *Critical Inquiry*, 15 (1988), 26–61 (pp. 31–7).

126 Robertson, *Mémoires*, I, 7, 112.

127 Smeaton, 'Some Large Burning Lenses', pp. 275–6.

128 See *Biographie universelle et portative*, 5 vols (Paris: Levraut, 1834), IV, 1122.

129 See David Francis Taylor, 'Gillray's Gulliver and the 1803 Invasion Scare', in *The Afterlives of Eighteenth-Century Fiction*, eds. Daniel Cook and Nicholas Seager (Cambridge: Cambridge University Press, 2015), pp. 212–32.

130 A. M. Broadley suspects that Gillray may have had a hand in this cartoon: see *Napoleon in Caricature, 1795-1821*, 2 vols (London: John Lane, The Bodley Head, 1911), I, 192.

131 Anon., 'A British Chymist Analizing a Corsican Earth Worm!' (London, July 1803), BM 1868,0808.7161.

132 Anon, 'Vent contraire' (Paris[?], November–December 1803), BM 1868,0808.6807.

133 See MacLeod, *Heroes of Invention*, p. 73.

134 Brutus, 'The Destruction of France by Mirrors', *The Belfast Monthly Magazine* 6: 34 (31 May 1811), 373–4. The author may have been at least inspired by, if not responsible for, the historical piece on burning-mirrors which featured in 1: 3 (1 November 1808), 183–5.

135 See Jonathan Sachs, *Romantic Antiquity: Rome in the British Imagination, 1789-1832* (Oxford: Oxford University Press, 2010), esp. p. 16.

136 Thomas Malthus, *An Essay on the Principle of Population* (London, 1798), p. 2.

137 The letter was published just days after the Duke had been re-instated as Commander-in-Chief. See H. M. Stephens, rev. John Van der Kiste, 'Frederick, Prince, duke of York and Albany', *Oxford Dictionary of National Biography*, accessed via <https://www.oxforddnb.com>.

138 Edmund Burke, *Reflections on the Revolution in France* (London, 1790), p. 8.

139 See Humphry Davy, 'Some Experiments on the Combustion of the Diamond and Other Carbonaceous Substances', *PT*, 104 (1814), 557–70 (pp. 557–8).

140 See R. G. W. Anderson, 'Joseph Black and His Chemical Furnace', in *Making Instruments Count: Essays on Historical Scientific Instruments presented to Gerard L'Estrange Turner*, eds. Anderson, J. A. Bennett and W. F. Ryan (Aldershot: Variorum, 1993), pp. 119–26.

Chapter 3

1 Robert Woof, 'Haydon, Benjamin Robert', *Oxford Dictionary of National Biography*, accessed via <https://www.oxforddnb.com>.

2 Willard Bissell Pope, ed., *The Diary of Benjamin Robert Haydon*, 5 vols (Cambridge, MA: Harvard University Press, 1960–3), I, 479.

3 On the fantasy of combustive annihilation, see Gaston Bachelard, *Psychoanalysis of Fire*, trans. Alan C. M. Ross (London: Routledge & Kegan Paul, 1964), p. 17.

4 Friedrich Schiller, 'An die Freude', ll. 49–50, in *Selected Poems*, ed. Frank M. Fowler (London: Macmillan, 1969), p. 3. Beethoven did not include these lines in his Symphony No. 9 in D Minor.

5 Wordsworth, *The Thirteen-Book Prelude*, ed. Mark L. Reed, 2 vols (Ithaca, NY: Cornell University Press, 1991), I, 131 (II, 273).

6 See esp. Jonathan Bate, *The Song of the Earth* (London: Picador, 2000).

7 *Memoirs of Goethe: Written by Himself*, 2 vols (London: for Henry Colburn, 1824), I, 27, 28, 29–30.

8 'pontifex, n', *OED Online*, accessed via <http://dictionary.oed.com>.

9 'Prometheus', ll. 21–7, in *The Complete Works of Johann Wolfgang von Goethe*, 10 vols (New York: Collier, 1839), V, 172–3.

10 *Goethe on Science: An Anthology of Goethe's Scientific Writings*, ed. Jeremy Naydler (Edinburgh: Floris Books, 1996), p. 128.

11 Goethe, *Scientific Studies*, ed. and trans. Douglas Miller (New York: Suhrkamp, 1988), p. 164. See Astrida Orle Tantillo, 'The Subjective Eye: Goethe's *Farbenlehre* and *Faust*', in *The Enlightened Eye: Goethe and Visual Culture*, eds. Evelyn K. Moore and Patricia Anne Simpson (New York: Rodopi, 2007), pp. 265–77 (p. 269).

12 *Conversations of Goethe with Eckermann and Soret*, ed. John Oxenford (London: Bell, 1875), p. 55 (2 January 1824).

13 Goethe, *Faust*, trans. Albert G. Latham, 2 vols (London: Dent, 1908), I, 15 (Part II, Act I); Arthur Zajonc, *Catching the Light: The Entwined History of Light and Mind* (Oxford: Oxford University Press, 1993), p. 215.

14 Goethe, *Sämtliche Werke. Jubiläumsausgabe in 40 Bänden*, ed. Eduard von
    der Hellen (Stuttgart, 1902–7), XXXVI, 115–16, translated in Rene Wellek, *A
    History of Modern Criticism, 1750-1959*, 5 vols (London: Jonathan Cape, 1955,
    repr. 1966), I, 203.

15 See Bettina von Arnim, *Goethe's Correspondence with a Child* (Boston: Osgood,
    1872), p. 493.

16 See esp. 'Walking with God', 'The Light and Glory of the Word' and 'The
    Shining Light' (*Olney Hymns*, I, XXX, and XXXII), and 'Charity' (ll. 589–98), in
    The Poems of William Cowper, eds. John D. Baird and Charles Ryskamp, 3 vols
    (Oxford: Clarendon Press, 1980–95), I, 139, 170, 172, 352.

17 M. H. Abrams, *The Mirror and the Lamp: Romantic Theory and the Critical
    Tradition* (New York: Oxford University Press, 1953), p. 57.

18 Priestley, *A Course of Lectures on Oratory and Criticism* (London, 1777), p. 31;
    Coleridge, 'On Poesy or Art', in *Biographia Literaria* (1817), ed. John Shawcross,
    2 vols (Oxford: Clarendon Press, 1907), II, 257–8.

19 Wordsworth, *The Thirteen-Book Prelude*, I, 133 (II, 387). The Romantics'
    expressive burning-glass had a long afterlife in models of cognition and
    feeling. See, for instance, John Ruskin, 'Imagination Contemplative', in
    The Works of John Ruskin, eds. E. T. Cook and Alexander Wedderburn,
    39 vols (London: George Allen and Unwin, 1903–12), IV, 290, Virginia
    Woolf, 'Sketch of the Past', in *Moments of Being*, ed. Jeanne Schulkind,
    rev. Hermione Lee (London: Pimlico, 2002), p. 103, and Hermione Lee, 'A
    Burning Glass: Reflection in Virginia Woolf', in *Virginia Woolf: A Centenary
    Perspective*, ed. Eric Warner (Basingstoke: Macmillan, 1984), pp. 12–27
    (pp. 16, 14).

20 See Ronald Paulson, *Representations of Revolution (1789-1820)* (New Haven:
    Yale University Press, 1983), pp. 45–6.

21 Marquis de Sade, 'Idées sur le mode de la sanction des lois' (November 1792),
    quoted in Chantal Thomas, *Sade* (Paris: Seuil, 1994), p. 181.

22 Mary Wollstonecraft, *An Historical and Moral View of the Origin and Progress of
    the French Revolution* (London, 1794), p. 12; Humphry Davy, 'The Progress of
    Reason – A Poem', Royal Institution MS HD/21/a (1795–June 1797), p. 23; Mary
    Robinson, *The Progress of Liberty*, I, 22–3, in *The Poetical Works of the Late
    Mrs Mary Robinson*, 3 vols in 1 (London, 1824), p. 164. See also Wollstonecraft,
    A Vindication of the Rights of Men (London, 1790), p. 66.

23 See esp. Stuart Curran, 'The Political Prometheus', *Studies in Romanticism*, 25:
    3 (Fall 1986), 429–55, and Linda M. Lewis, *The Promethean Politics of Milton,
    Blake, and Shelley* (Columbia: University of Missouri Press, 1992).

24 See Carol Dougherty, *Prometheus* (London: Routledge, 2006), *passim*, and Olga Raggio, 'The Myth of Prometheus: Its Survival and Metamorphoses Up to the Eighteenth Century', *Journal of the Warburg and Courtauld Institutes*, 21: 1/2 (January–June 1958), 44–62.

25 See Raphael Thorius, *Hymnus tabaci*, trans. Peter Hausted (1626; London, 1651), p. 19, Charles Sorel, *The Extravagant Shepherd, the Anti-Romance*, trans. anon (London, 1653), p. 65, and Athanasius Kircher, *Ars Magna Lucis et Umbræ*, 2nd edn (Amsterdam, 1671), p. 424.

26 Richard Brinsley Sheridan, *Pizarro: A Tragedy* (London, 1799) pp. 21–4 (Act II, Scene II). Byron's poem also brings to mind Chaucer's (and Pope's) glass temple of Fame.

27 Byron, 'Monody on the Death of the Right Hon. R. B. Sheridan, Spoken at Drury Lane Theatre, London', in *Lord Byron: The Complete Poetical Works*, ed. Jerome J. McGann, 7 vols (Oxford: Clarendon Press, 1980–93), IV, 19–20. See also *Don Juan*, Canto II, ll. 1482–3, in *Complete Poetical Works*, V, 146–7.

28 See Michael Simpson, 'On Byron's Famous Fanes: Ruined Temples and Reformed Theatres', *Byron Journal*, 41: 2 (2013), 145–57 (p. 154).

29 See Wordsworth, 'Lines, Composed at Grasmere' (pub. 1807), ll. 17–18, in *William Wordsworth*, ed. Stephen Gill (Oxford: Oxford University Press, 1984), p. 329.

30 Robert Smith, *The Elementary Parts of Dr. Smith's Compleat System of Opticks* (Cambridge, 1778), p. 9.

31 William Herschel, 'Investigation of the Powers of the Prismatic Colours to Heat and Illuminate Objects; with Remarks, That Prove the Different Refrangibility of Radiant Heat', *Philosophical Transactions* (hereafter *PT*), 90 (1800), 255–83 (p. 255).

32 The variation in heat intensity across the visible spectrum had been noticed previously, but not mapped systematically. See E. S. Cornell, 'The Radiant Heat Spectrum from Herschel to Melloni-I. The Work of Herschel and his Contemporaries', *Annals of Science*, 3 (1938), 119–37 (p. 119).

33 Herschel, 'Investigation of the Powers of the Prismatic Colours', p. 272.

34 Herschel, 'Experiments on the Refrangibility of the Invisible Rays of the Sun', *PT*, 90 (1800), 284–92 (pp. 292, 291).

35 Herschel, 'Experiments on the Solar, and on the Terrestrial Rays That Occasion Heat; […] Part I', *PT*, 90 (1800), 293–326. In 1790, Marc-Auguste Pictet had shown that parabolic mirrors could reflect cold: see *An Essay on Fire*, trans. W. B[elcombe] (London, 1791), pp. 116–23.

36 Herschel, 'Experiments on the Solar […] Part II', pp. 490, 507; Joseph Banks to Herschel, 9 April 1800, Royal Astronomical Society MSS, B. 34, ff. 1^v-2^r. On Rumford's work, see esp. 'An Enquiry concerning the Nature of Heat, and the Mode of its Communication', *PT*, 94 (January 1804), 77–182.

37 Barbauld, 'Eternity' (c.1809–1810), ll. 18–20, in *The Poems of Anna Letitia Barbauld*, eds. William MacCarthy and Elizabeth Kraft (Athens: The University of Georgia Press, 1994), p. 149.

38 Banks to Herschel, 19 March 1800, Royal Astronomical Society MSS, B. 32, f. 1^r; *Prometheus Unbound*, IV, 230, in *The Poems of Shelley*, eds. Michael Rossington and others, 5 vols (London: Routledge, 1989–2014), II, 626. See Carl Grabo, *A Newton Amongst Poets: Shelley's Use of Science in 'Prometheus Unbound'* (Chapel Hill: University of North Carolina Press, 1930), p. 111.

39 Banks to Herschel, 19 March 1800, Royal Astronomical Society MSS, B. 32, f. 1^r.

40 Herschel to Banks, 20 March 1800, in *The Scientific Correspondence of Sir Joseph Banks*, V, 31.

41 Banks to Herschel, 24 March 1800, in *Scientific Correspondence*, V, 34. The caloric theory took many years to lose its popularity, however: see Robert Fox, *The Caloric Theory of Gases: From Lavoisier to Regnault* (Oxford: Clarendon Press, 1971), esp. pp. 104–5.

42 See John Leslie, 'Observations and Experiments on Light and Heat, with some Remarks on the Enquiries of Dr Herschel, Respecting Those Objects', *A Journal of Natural Philosophy, Chemistry, and the Arts*, ed. William Nicholson, 4 (1800), 344–50.

43 Davy to Davies Giddy, 14 November 1801, in John Ayrton Paris, *The Life of Sir Humphry Davy*, 2 vols (London: Colburn and Bentley, 1831), I, 86.

44 See Walter D. Wetzels, 'Johann Wilhelm Ritter: Romantic Physics in Germany', in *Romanticism and the Sciences*, eds. Andrew Cunningham and Nicholas Jardine (Cambridge: Cambridge University Press, 1990), pp. 199–212 (pp. 207–8).

45 Thomas Young, 'The Bakerian Lecture: On the Theory of Light and Colours', *PT*, 92 (1802), 12–48 (p. 47). On Herschel's commitment to projectile theory, see Cantor, *Optics after Newton*, p. 169.

46 See C. J. Wright, 'The "Spectre" of Science: The Study of Optical Phenomena and the Romantic Imagination', *Journal of the Warburg and Courtauld Institutes*, 43 (1980), 186–200 (p. 199).

47 See Marina Warner, 'Spirit Visions', in *The Tanner Lectures on Human Values* (Yale, 1999), p. 84.

48 Shelley, *Prometheus Unbound*, II.iv.3-4, in *The Poems of Shelley*, II, 556; H. G. Wells, *The War of the Worlds*, eds. David Y. Hughes and Harry M. Geduld (Bloomington: Indiana University Press, 1993), p. 66.

49 Mary Shelley, *Frankenstein: The 1818 Text*, ed. J. Paul Hunter (New York: Norton, 1996), p. 28 (Vol. I, Chap. II).

50 See, for instance, the *Ancient Mariner*'s epigraph from Thomas Burnet, in *The Collected Works of Samuel Taylor Coleridge*, general ed. Kathleen Coburn, 16 vols (Princeton, NJ: Princeton University Press, 1969–2002), XVI Part 1, I.I, 371.

51 See Jn. 9. 4-5; 12. 35-6; 1 Cor. 4. 5.

52 Wordsworth, *The Thirteen-Book Prelude*, I, 316 (XIII, 104–5), I, 190 (VI, 534–6). See also Coleridge, 'To William Wordsworth, Composed on the Night After his Recitation of a Poem on the Growth of an Individual Mind' (1807), ll. 19–20, in *Collected Works*, XVI Part 1, I.II, 816.

53 Gillian Beer, '"Authentic Tidings of Invisible Things": Vision and the Invisible in the Later Nineteenth Century', in *Vision in Context: Historical and Contemporary Perspectives on Sight*, eds. Teresa Brennan and Martin Jay (New York: Routledge, 1996), pp. 85–94 (p. 86).

54 See Davy to Davies Giddy, 3 July 1800, in Paris, *The Life of Sir Humphry Davy*, I, 59.

55 Coleridge to Davy, 4 May 1801, in *Collected Letters of Samuel Taylor Coleridge*, ed. E. L. Griggs, 6 vols (Oxford: Clarendon Press, 1956–71), I, 727. On the projected laboratory, see Coleridge to Davy, 3 February 1801, in *Collected Letters*, I, 670.

56 See Paris, *The Life of Sir Humphry Davy*, I, 92, and *The Notebooks of Samuel Taylor Coleridge*, eds. Kathleen Coburn and Anthony John Harding, 5 vols in 10 (London: Routledge & Kegan Paul, 1959–2002), I, 1098–9. On Davy's profound influence on Coleridge, see Trevor H. Levere, *Poetry Realized in Nature: Samuel Taylor Coleridge and Early Nineteenth-Century Science* (Cambridge: Cambridge University Press, 1981), esp. pp. 23–35.

57 Coleridge, *Notebooks*, I, 1233. See also *The Friend* (1818), Vol. I, Essay XIV, in *Collected Works*, IV.I, 105.

58 Coleridge, 'Hymn before Sun-rise, in the Vale of Chamouni', in *Collected Works*, XVI Part 1, I.II, 720 (ll. 1, 2).

59 Coleridge, *Lectures on Literature*, 2 vols (Vol 5 of *Collected Works*), II, 220.

60 Coleridge, *Aids to Reflection*, in *Collected Works*, IX, 398.

61 Coleridge to Thomas Poole, 5 November 1796, in *Collected Letters*, I, 250.

62 Coleridge, *Notebooks*, III, 3377. On fire images in Coleridge's other Asra poems, see George Whalley, *Coleridge and Sara Hutchinson and the Asra Poems* (London: Routledge & Paul, 1955), pp. 127–8.

63 See Coleridge, *Notebooks*, II, 2984 (where he compares love to a silent, subterranean volcano).

64   See Coleridge, *Notebooks*, II, 3148.

65   See Dante Alighieri, *The Divine Comedy*, trans. John D. Sinclair, 3 vols (1939; London: Oxford University Press, 1971), I, 78 (*Inferno*, Canto V).

66   Coleridge, *The Friend*, Vol. I, Essay XIII, in *Collected Works*, IV.I, 94n.

67   See Coleridge, *Notebooks*, III, 3379n; I, 1233; II, 3222.

68   Humphry Davy, 'An Essay on Heat, Light, and the Combinations of Light', in *Contributions to Physical and Medical Knowledge*, ed. Thomas Beddoes (Bristol, 1799), pp. 5–147 (pp. 8, 43, 64, 60). On the evolution of Davy's thought on light, see Davy to Davies Giddy, 22 February 1799, in *Memoirs of the Life of Sir Humphry Davy, Bart.*, ed. John Davy, 2 vols (London: for Longman, 1836), I, 76–9.

69   On Davy's belief in light as the principle of life, see Sharon Ruston, *Creating Romanticism: Case Studies in the Literature, Science and Medicine of the 1790s* (Houndmills: Palgrave Macmillan, 2013), esp. pp. 164–5.

70   See Jan Ingenhousz, *Experiments upon Vegetables, Discovering Their Great Power of Purifying the Common Air in the Sun-shine* (London, 1779).

71   Davy, 'A Syllabus of A Course of Lectures of Chemistry, Delivered at the Royal Institution of Great Britain' (1802), in *The Collected Works of Sir Humphry Davy, Bart.*, ed. John Davy, 9 vols (London: Smith, Elder and Co, 1839–40), II, 388.

72   Davy, 'A Syllabus', in *Works*, II, 393.

73   See esp. Alice Jenkins, 'Humphry Davy: Poetry, Science and the Love of Light', in *1798: The Year of 'Lyrical Ballads'*, ed. Richard Cronin (Houndmills: Macmillan, 1998), pp. 133–50, and Maurice Hindle, 'Nature, Power, and the Light of Suns: The Poetry of Humphry Davy', *The Charles Lamb Bulletin*, new series 157 (Spring 2013), 38–54.

74   *Memoirs of the Life of Sir Humphry Davy, Bart.*, I, 82. On Davy's theology of light, see J. Z. Fullmer, 'The Poetry of Sir Humphry Davy', *Chymia*, 6 (1960), 102–26 (p. 108).

75   See Maurice Hindle, 'Humphry Davy and William Wordsworth: A Mutual Influence', *Romanticism*, 18: 1 (2012), 16–29 (p. 17).

76   Davy, 'Written after Recovery from a Dangerous Illness' (1808), in *Collected Works*, I, 114.

77   James Hutton, *A Dissertation upon the Philosophy of Light, Heat, and Fire* (Edinburgh, 1794), p. 69. See Ted Underwood, *The Work of the Sun: Literature, Science, and Economy, 1760-1860* (New York: Palgrave Macmillan, 2005), p. 69.

78   Wordsworth, 'Lines Written a Few Miles above Tintern Abbey', l. 98, in *Lyrical Ballads, and Other Poems, 1797-1800*, eds. James Butler and Karen Green (Ithaca, NY: Cornell University Press, 1992), p. 119.

79 Davy, 'The life of the Spinosist', in Royal Institution Davy MSS, Notebook
    13c, 9.

80 Davy, 'Ullswater August 4 1825', in *Collected Works*, I, 321. See also Royal
    Institution Davy MSS, Notebook 14e, 97.

81 *Fragmentary Remains, Literary and Scientific, of Sir Humphry Davy, Bart.*, ed.
    John Davy (London: John Churchill, 1858), p. 55.

82 Davy, 'A Discourse Introductory to a Course of Lectures on Chemistry,
    Delivered in the Theatre of the Royal Institution, on the 21st of January, 1802',
    in *Collected Works*, II, 307–26 (pp. 318–19). See Ruston, *Creating Romanticism*,
    pp. 172–3.

83 See esp. Laura E. Crouch, 'Davy's *A Discourse, Introductory to A Course of
    Lectures on Chemistry*: A Possible Scientific Source of *Frankenstein*', *Keats-
    Shelley Journal*, 27 (1978), 35–44.

84 See *The Journals of Mary Shelley: 1814-1844*, eds. Paula R. Feldman and Diana
    Scott-Kilvert, 2 vols (Oxford: Clarendon Press 1987), I, 143 (29 October 1816).

85 Davy, 'An Essay on Heat', p. 127; Davy, 'A Discourse, Introductory', in *Collected
    Works*, II, 321; *Frankenstein*, pp. 30, 32 (Vol. I, Chap. III).

86 *Curiosity Perfectly Satisfied: Faraday's Travels in Europe 1813-1815*, eds. Brian
    Bowers and Lenore Symons (London: Peter Peregrinus, in association with the
    Science Museum, 1991), p. 73 (21 March 1814).

87 See Davy to Banks, 21 March 1814, in *The Scientific Correspondence of Sir
    Joseph Banks*, VI, 129–30, and Davy to Alexander Marcet, 29 July 1814, accessed
    via <www.davy-letters.org.uk>.

88 See Smithson Tennant, 'On the Nature of the Diamond', *PT*, 87 (1797), 123–7,
    and Robert Siegfried, 'Sir Humphry Davy on the Nature of the Diamond', *Isis*,
    57: 3 (Autumn 1966), 325–35 (p. 328).

89 Davy, *Elements of Chemical Philosophy* (London: for J. Johnson, 1812), p. 312.

90 *Curiosity Perfectly Satisfyed*, p. 75 (27 March 1814).

91 *Curiosity Perfectly Satisfyed*, p. 76 (28 March 1814).

92 Faraday to Benjamin Abbott, 1 May and 24 July 1814, in *The Correspondence
    of Michael Faraday*, ed. Frank A. J. L. James, 6 vols (London: Institution of
    Electrical Engineers, 1991), I, 76 (Letter 33). For discussion of the Buffon lens
    between Faraday, Georges-Mathilde-Ernest Degrand and Thomas Stevenson,
    see *Correspondence*, V, 587–91, 599, 601–5 (Letters 3673, 3682 and 3687
    (all 1859)).

93 See Davy, 'Some Experiments on the Combustion of the Diamond and Other
    Carbonaceous Substances', 557–70 (p. 568), and Davy to Banks, 1 May 1814,
    accessed via <www.davy-letters.org.uk>.

94 Indeed, in a poem reworked in 1816/17 as 'On the Immortality of the Mind',
    'intellectual light' is celebrated as 'burning still / Its lustre purer and more
    bright' (Davy, *Collected Works*, I, 235). Alice Jenkins notes that, rather than
    'burning', '"shining" might have been a more obvious choice' ('Humphry Davy',
    p. 146). Davy's interest in combustion is also evident in 'Aug 1. Copenhagen 24',
    in Royal Institution Davy MSS, Notebook 14e, 100.

95 William Hamilton Drummond, *The Giant's Causeway, A Poem* (Belfast:
    Longman and others, 1811), p. 74 (III, 58, 62–5).

96 Honoré de Balzac, *The Alkahest: Or, The House of Claës*, ed. Katharine Prescott
    Wormeley (Boston: Roberts Brothers, 1890), p. 109. On this novel's broader
    relationship with chemistry, see Sharon Ruston, 'Chemistry', in *The Routledge
    Companion to Nineteenth-Century British Literature and Science*, eds. John
    Holmes and Sharon Ruston (Oxford: Routledge, 2017), pp. 271–85 (esp. p.
    278), and John Tresch, 'Electromagnetic Alchemy in Balzac's *The Quest for
    the Absolute*', in *The Shape of Experiment*, eds. Henning Schmidgen and Julia
    Kursell (Berlin: Max-Planck, 2007), pp. 57–77.

97 George Tucker, *A Century Hence: Or, A Romance of 1941*, ed. Donald R. Noble
    (Charlottesville: University Press of Virginia, 1997), p. 68. Tucker's visions
    are perhaps satirizing the techno-utopian John Etzler's *The Paradise Within
    the Reach of All Men* (Pittsburgh: Etzler and Reinhold, 1833), which includes
    burning-mirrors (pp. 34–46).

98 See esp. Nora Crook, 'Shelley and the Solar Microscope', *Keats-Shelley Review*, 1
    (1986), 49–60.

99 Anon (probably John Taylor Coleridge), 'Review of Leigh Hunt, *Foliage*',
    Quarterly Review, 18: 36 (January 1818), 324–35 (p. 327n).

100 Thomas Jefferson Hogg, *The Life of Percy Bysshe Shelley*, 4 vols (London:
    Edward Moxon, 1858), I, 41–2, 40. John Moultrie immortalized Shelley's
    pyromania poetically: see the stanzas quoted in H. C. Maxwell Lyte, *A History of
    Eton College, 1440-1875* (London: Macmillan, 1875), p. 398.

101 A. A. [Andrew Amos], 'Shelley and His Contemporaries at Eton', *The Athenæum:
    The Journal of English and Foreign Literature, Science, and the Fine Arts*, Issue
    1058 (15 April 1848), 390–1 (p. 390). Thomas Medwin similarly noted that his
    cousin Shelley was 'very superficial' in chemistry: *The Shelley Papers: Memoir of
    Percy Bysshe Shelley* (London: Whittaker, Treacher, & Co, 1833), p. 13.

102 Adam Walker, *A System of Familiar Philosophy: In Twelve Lectures* (London,
    1799), pp. 2, 10. See also p. 11. As Sharon Ruston points out, it is possible
    Shelley read a later edition: see *Shelley and Vitality* (Houndmills: Palgrave
    Macmillan, 2005), p. 190n25.

103   Percy Shelley to Thomas Hookham, 29 July 1812, in *The Letters of Percy Bysshe Shelley*, ed. Frederick L. Jones, 2 vols (Oxford: Clarendon Press, 1964), I, 319; Davy, *Elements of Chemical Philosophy*, pp. 203–4.

104   See Shelley, *Queen Mab* (1813), IX, 224 (*Poems*, I, 359), revised in *The Daemon of the World: A Fragment* (1815), II, 316 (*Poems*, I, 508). See also *Queen Mab*, I, 242–3 (*Poems*, I, 360).

105   In Shelley's *Epipsychidion* (1821) (ll. 163–9) and 'Ode to Liberty' (1820) (l. 138), respectively, imagination and liberty are figured sending 'sun-like arrow[s]' and 'shafts' to destroy 'error' (*Poems*, IV, 145–6; III, 403).

106   *The Works of Percy Bysshe Shelley in Verse and Prose*, ed. Harry Buxton Forman, 8 vols (London: Reeves and Turner, 1880), VII, 153, and III, 344 (l. 351). Cf. Shelley to Charles Ollier, early March 1821, in *Letters*, II, 273.

107   'Lines Written in a Blank Leaf of the "Prometheus Unbound."' (1822), ll. 1–2, in *The Poetical Works of Thomas Lovell Beddoes*, ed. Edmund Gosse, 2 vols (London: Dent, 1890), I, 27; 'Pauline: A Fragment of a Confession' (1833), ll. 151, 171, in *The Poetical Works of Robert Browning*, eds. Ian Jack and Margaret Smith, 15 vols (Oxford: Clarendon Press, 1983–2009), I, 34, 36.

108   See M. H. Abrams, *Natural Supernaturalism: Tradition and Revolution in Romantic Literature* (London: Oxford University Press, 1971), p. 307.

109   As Timothy Morton suggests, this techno-utopian 'machine of light' is associated with the progress of liberty: see 'Shelley's Green Desert', *Studies in Romanticism*, 35: 3 (1996), 409–30 (pp. 411, 417).

110   See Cantor, *Optics after Newton*, pp. 61–2, 156–8.

111   Shelley, *Queen Mab*, IX, 152–3; *Poems*, I, 357.

112   See Ruston, *Shelley and Vitality*, pp. 32, 117.

113   See Grabo, *A Newton Amongst Poets*, pp. 144–8, and Carl Grabo, *Prometheus Unbound: An Interpretation* (Chapel Hill: University of North Carolina Press, 1935), pp. 142–3. See also Shelley, *The Cenci* (1819), I.ii.84–5 (*Poems*, II, 749), in which the lustful Prelate Orsino imagines the 'anatomiz[ing]' gaze of Beatrice.

114   Keats to Fanny Keats, 28 August 1819, in *The Letters of John Keats*, ed. Maurice Buxton Forman, 3rd edn (London: Oxford University Press, 1947), p. 375.

115   Keats to George and Georgiana Keats, 18 September 1819, in *Letters*, p. 402.

116   Keats, *Lamia*, I, 8, in *The Poems of John Keats*, ed. Miriam Allott (London: Longman, 1970), p. 617.

117   See Stuart M. Sperry, Jr., 'Keats and the Chemistry of Poetic Creation', *PMLA*, 85: 2 (March 1970), 268–77 (pp. 275–6), Denise Gigante, 'The Monster in the Rainbow: Keats and the Science of Life', *PMLA*, 117: 3 (May 2002), 433–48

(p. 439), and Dometa Wiegand Brothers, *The Romantic Imagination and Astronomy: On All Sides Infinity* (Basingstoke: Palgrave Macmillan, 2015), pp. 112–14.

118 *A Syllabus of a Course of Chemical Lectures Read at Guy's Hospital, by William Babington, Alexander Marcet and William Allen* (London: by William Phillips, 1816), p. 9. These lectures probably informed Oceanus's description of the generative power of light in *Hyperion*, II, 195–7 (*The Poems of John Keats*, p. 427).

119 The most famous appropriation of the phrase is probably Richard Dawkins, *Unweaving the Rainbow: Science, Delusion and the Appetite for Wonder* (London: Allen Lane, 1998), p. xii. Keats's original was 'Destroy a rainbow': see *Poetry Manuscripts at Harvard: A Facsimile Edition*, ed. Jack Stillinger (Cambridge, MA: The Belknap Press of Harvard University Press, 1990), p. 215.

120 Byron, 'Darkness', ll. 2, 72, in *Complete Poetical Works*, IV, 40, 43. On Byron's experience of the poor weather, see Byron to Samuel Rogers, 28 July 1816, in *Byron's Letters and Journals*, ed. Leslie A. Marchand, 13 vols (London: Murray, 1973–94), V, 86. On 'Darkness' and Tambora, see esp. Anthony Rudolf, *Byron's Darkness: Lost Summer and Nuclear Winter* (London: The Menard Press, 1984), pp. 3–5, and Bate, *The Song of the Earth*, pp. 94–8. On 'Darkness' and sunspots, see Jeffrey Vail, '"the Bright Sun was Extinguis'd": The Bologna Prophecy and Byron's "Darkness"', *The Wordsworth Circle*, 28: 3 (Summer 1997), 183–92.

121 *The Poetical Works of Thomas Campbell*, 2 vols (London: Edward Moxon, 1837), II, 107.

Chapter 4

1 *The Poetical Works of Anna Seward*, ed. Walter Scott, 3 vols (Edinburgh: Ballantyne, 1810), III, 184. See also the longer blank verse poem 'Colebrook Dale' (*Poetical Works*, II, 314–19). Seward celebrates the Sun as the bringer of vitality: see esp. 'Ode to the Sun' (1780) and 'Address to the Sun' (1782), in *Poetical Works*, II, 49–52, 174–5. In several poems, Seward figures Shakespeare's genius as a 'solar flame': see, for example, 'Prologue Written for Mr Penn' (1782), l. 32 (*Poetical Works*, I, 169).

2 William Stanley Jevons, *The Coal Question: An Inquiry Concerning the Progress of the Nation, and the Probable Exhaustion of Our Coal-mines*, 2nd edn, revised (London: Macmillan, 1866), p. 2.

3 Frederick Soddy, *Wealth, Virtual Wealth, and Debt*, 2nd edn (London: George Allen, 1933), p. 29.

4 See J. Homer Lane, 'On the Theoretical Temperature of the Sun' (1870), in *Early Solar Physics*, ed. A. J. Meadows (Oxford: Pergamon Press, 1970), pp. 257–76.

5 Balfour Stewart and J. Norman Lockyer, 'The Sun as a Type of the Material Universe', Part I, *Macmillan's Magazine*, 18 (July 1868), 246–57 (p. 249).

6 See Mike Hulme, 'On the Origin of "the Greenhouse Effect": John Tyndall's 1859 Interrogation of Nature', *Weather*, 64: 5 (May 2009), 121–3.

7 See Stuart Peterfreund, 'The Re-Emergence of Energy in the Discourse of Literature and Science', *Annals of Scholarship*, 4 (1986–7), 22–53 (pp. 24–5).

8 See Gillian Beer, '"The Death of the Sun": Victorian Solar Physics and Solar Myth', in *The Sun Is God: Painting, Literature and Mythology in the Nineteenth Century*, ed. J. B. Bullen (Oxford: Clarendon Press, 1989), pp. 159–80, repr. in *Open Fields: Science in Cultural Encounter* (Oxford: Oxford University Press, 1996), pp. 219–41.

9 On 'thermodynamic optimism', see Barri J. Gold, *Thermopoetics: Energy in Victorian Literature and Science* (Cambridge, MA: MIT Press, 2010), esp. p. 30.

10 Anon., 'Proctor's Poetry of Astronomy', *The Saturday Review*, 28 May 1881, 693–4 (p. 693) (a review of Richard A. Proctor, *The Poetry of Astronomy: A Series of Familiar Essays on the Heavenly Bodies, &c.* (London: Smith, Elder & Co, 1881)).

11 Felicia Hemans, 'The Sunbeam', in *The Poetical Works of Mrs. Felicia Hemans; Complete in One Volume* (Philadelphia: Grigg & Elliot, 1841), p. 283.

12 See Bullen, ed., *The Sun Is God*, and esp. Dinah Birch's chapter '"The Sun Is God": Ruskin's Solar Mythology', pp. 109–23.

13 Thomas De Quincey, *Suspiria de Profundis* (1845), in *Confessions of an English Opium-Eater and Other Writings*, ed. Barry Milligan (London: Penguin, 2003), p. 89.

14 See Geoffrey Batchen, *Burning with Desire: The Conception of Photography* (Cambridge, MA: MIT Press, 1999), esp. pp. 26–30, 63, 101.

15 See Alex Soojung-Kim Pang, 'Victorian Observing Practices, Printing Technology, and Representations of the Solar Corona (1): The 1860s and 1870s', *Journal of the History of Astronomy*, 25 (1994), 249–74 (p. 252).

16 See, for instance, J. Norman Lockyer, *Contributions to Solar Physics* (London: Macmillan & Co, 1874), after pp. 240 and 374.

17 Jules Janssen, 'Notes on Recent Progress in Solar Physics' (1879), in *Early Solar Physics*, pp. 135–68 (p. 153).

18 See, for instance, Anon., 'On Burning Glasses', *Saturday Magazine*, 16: 512 (27 June 1840), 245–7, and John Henry Pepper, *The Boy's Play-book of Science*, new edn (London: Routledge, 1869), pp. 291–4.

19 See Thomas Stevenson to Michael Faraday, 19 November 1859, in *The Correspondence of Michael Faraday*, V, 599 (Letter 3682), and John Perlin, *Let It Shine: The 6,000-Year Story of Solar Energy* (Novato, CA: New World Library, 2013), pp. 74–7.

20 See Charles Dickens, *David Copperfield*, ed. Nina Burgis (Clarendon Press, 1981), p. 238 (Chapter XIX), Dickens, *American Notes For Circulation* (London: Chapman and Hall, 1874), p. 91 (Chap. 6), Dickens, *The Uncommercial Traveller and Other Papers, 1859–70*, eds. Michael Slater and John Drew (Columbus: Ohio State University Press, 2000), p. 87, George Eliot to Maria Lewis, 4 September 1839, in *The George Eliot Letters*, ed. Gordon S. Haight, 9 vols (New Haven: Yale University Press, 1954–78), I, 29, Eliot, *Daniel Deronda*, ed. Graham Handley (Oxford: Clarendon Press, 1984), p. 250 (III. xxiv), Eliot, *The Mill on the Floss* (London: The Zodiac Press, 1951), p. 72 (I. vii), and Thomas Hardy, *Two on a Tower*, ed. Suleiman M. Ahmad (Oxford: Oxford University Press, 1993), esp. pp. 221, 240.

21 Charlotte Brontë, *Jane Eyre*, eds. Jane Jack and Margaret Smith (Oxford: Clarendon Press, 1969), p. 73 (Vol. I, Chap. VII). For Jane's apostrophe to sunshine, see p. 415 (Vol. III, Chap. II).

22 On optics as a metaphor of mind more generally, see W. David Shaw, 'The Optical Metaphor: Victorian Poetics and the Theory of Knowledge', *Victorian Studies*, 23: 3 (1980), 293–344. Influenced by David Brewster, Brontë's *Villette* (1853) combines traditional emblems of the eye of love's piercing ray with ophthalmic innovations and (possibly) burning-glasses. See Heather Glen, *Charlotte Brontë: The Imagination in History* (Oxford: Oxford University Press, 2002), pp. 214–29, and Katherine Inglis, 'Ophthalmoscopy in Charlotte Brontë's *Villette*', *Journal of Victorian Culture*, 15: 3 (December 2010), 348–69 (esp. pp. 359–60).

23 See also Henry Ellison, 'On the Great Exhibition of 1851', ll. 10–11, in *Stones from the Quarry; Or, Moods of Mind* (London: Provost, 1875), p. 190.

24 See William Clyde DeVane, *A Browning Handbook* (London: John Murray, 1935), p. 445.

25 See esp. Roma A. King Jr., *The Focusing Artifice: The Poetry of Robert Browning* (Athens, OH: Ohio University Press, 1968), p. 239.

26 Robert Browning to John Ruskin, 10 December 1855, in *Robert Browning: The Critical Heritage*, eds. Boyd Litzinger and Donald Smalley (London: Routledge & Kegan Paul, 1970), p. 15.

27 Browning, 'Parleying with Bernard de Mandeville', in *The Poetical Works of Robert Browning*, eds. Ian Jack et al., 15 vols projected (Oxford: Clarendon Press, 1983-date), XV, 60, 61 (ll. 277–81, 298–306).

28 See also Browning's optical metaphors in his letter to Elizabeth Barrett, 13 January 1845, in *The Brownings' Correspondence*, eds. Philip Kelley and Scott Lewis, 25 vols to date (Winfield, KS: Wedgestone Press, 1992), X, 22.

29 See William O. Raymond, "'The Jewelled Bow": A Study in Browning's Imagery and Humanism', *PMLA*, 70 (1955), 115–31 (p. 124). See also Raymond, 'Browning and Higher Criticism', *PMLA*, 44 (1929), 590–621 (pp. 617–18).

30 Cf. Browning's *The Ring and the Book*, X, 1311–12, when the pope conceives of 'Man's mind [as] a convex glass / Wherein are gathered all the scattered points', fragmentary understandings of God (*Poetical Works*, IX, 157).

31 See Dorothy Mermin, 'Browning and the Primitive', *Victorian Studies*, 25: 2 (Winter 1982), 211–37 (p. 212).

32 F. Max Müller, *Lectures on the Origin and Growth of Religion* (London: Longmans, Green and Co., 1878), pp. 113, 207–8. On Browning's acquaintance with Müller, see Kingsbury Badger, "'See the Christ Stand!" Browning's Religion' (1955), in *Robert Browning: A Collection of Critical Essays*, ed. Phillip Drew (London: Routledge, 1966), pp. 72–95 (p. 76).

33 See Beer, *Open Fields*, p. 224.

34 See esp. Ivan Kreilkamp, "'One More Picture": Robert Browning's Optical Unconscious', *ELH*, 73 (2006), 409–35 (p. 423).

35 *Walpole to George Montagu, 15 June 1768*, in *Horace Walpole's Correspondence*, ed. W. S. Lewis, 48 vols (New Haven: Yale University Press, 1937–83), X, 262.

36 See *The Collected Works of Sir Humphry Davy, Bart.*, ed. John Davy, 9 vols (London: Smith, Elder and Co, 1839–40), IV, 172.

37 Samuel Smiles, *The Life of George Stephenson, Railway Engineer*, 2nd edn (London: John Murray, 1857), pp. 484–5.

38 See, for instance, William Buckland, *Geology and Mineralogy Considered with Reference to Natural Theology*, 2 vols (London: W. Pickering, 1836), I, 352.

39 Anon., 'The Place Where Light Dwelleth', *The British Quarterly Review*, 51 (April 1870), 409–41 (p. 440).

40 See Allen MacDuffie, *Victorian Literature, Energy and the Ecological Imagination* (Cambridge: Cambridge University Press, 2014), pp. 29–30.

41 Balfour Stewart and J. Norman Lockyer, 'The Sun as a Type of the Material Universe', Part II, *Macmillan's Magazine*, 18 (August 1868), 319–27 (p. 322).

42 See Anon., 'More Work for the Sun', *All the Year Round*, 16: 401 (5 August 1876), 490–3 (p. 492), and Anon., 'The Place Where Light Dwelleth', esp. pp. 413–14.

43 See MacDuffie, *Victorian Literature*, p. 49.

44 Jevons Papers, John Rylands Library, Manchester, JA6/9/168, quoted in Andreas Malm, *Fossil Capital: The Rise of Steam Power and the Roots of Global Warming* (London: Verso, 2016), p. 159.

45 Francis R. Upton, 'Edison's Electric Light', *Scribner's Monthly*, 19: 4 (4 February 1880), 531–44 (pp. 531, 544).

46 See Wolfgang Schivelbusch, *Disenchanted Night: The Industrialisation of Light in the Nineteenth Century* (Oxford: Berg, 1988), pp. 53, 120.

47 See Schivelbusch, *Disenchanted Night*, pp. 3, 128, 130.

48 John Tyndall, 'On Force', *Philosophical Magazine*, 24 (1862), 57–66 (p. 63).

49 See Jonathan Smith, 'Writing Science: Scientific Prose', in *The Routledge Companion to Nineteenth-Century British Literature and Science*, eds. John Holmes and Sharon Ruston (Oxford: Routledge, 2017), pp. 141–54 (pp. 141–2).

50 *Modern Painters*, Volume I, in *The Works of John Ruskin*, eds. E. T. Cook and Alexander Wedderburn, 39 vols (London: George Allen, 1903–12), III, 279.

51 On the importance of the Sun within thermodynamics, in bringing physics and biology together, see Bruce Clarke, *Energy Forms: Allegory and Science in the Era of Classical Thermodynamics* (Ann Arbor: The University of Michigan Press, 2001), p. 151.

52 John Tyndall, *Heat Considered as a Mode of Motion* (London: Longman, 1863), p. 433, and Tyndall, 'On Force', p. 64. Cf. William Thomson, 'On the Mechanical Action of Radiant Heat or Light', *Proceedings of the Royal Society of Edinburgh*, 3 (1857), 108–13 (p. 113).

53 Letter 63 ('Sit Splendor'), March 1876, in *The Works of John Ruskin*, XXVIII, 541. See also *Works*, XXVI, 183, and Francis O'Gorman, *Late Ruskin: New Contexts* (Aldershot: Ashgate, 2001), pp. 110–11.

54 MacDuffie, *Victorian Literature*, p. 40.

55 Balfour Stewart, *The Conservation of Energy* (New York: D. Appleton, 1875), p. 144.

56 Nathaniel Bagshaw Ward, *On the Growth of Plants in Closely Glazed Cases* (London: John Van Voorst, 1842), p. 11.

57 Newton, *Opticks*, p. 318 (Book III, Query 11).

58 Swift, *Gulliver's Travels*, p. 237 (Book III, Chapter II).

59 See Simon Schaffer, '"The Great Laboratories of the Universe": William Herschel on Matter Theory and Planetary Life', *Journal of the History of Astronomy*, 11 (1980), 81–111 (p. 83).

60 See, for instance, Jacob Berzelius's theory that solar energy production was analogous to an electric pile: 'An Explanatory Statement of the Notions or Principles upon which the Systematic Arrangement is founded, which was adopted as the Basis of an Essay on Chemical Nomenclature' (1811), repub. in *Journal of Natural Philosophy, Chemistry and the Arts*, 34 (1813), 142–6 (p. 144).

61 See Peterfreund, 'The Re-Emergence of Energy', and Helge Kragh, 'The Source of Solar Energy, ca. 1840–1910: From Meteoric Hypothesis to Radioactive Speculations', *The European Physical Journal*, 41 (2016), 365–94.

62 See P. M. Harman, *Energy, Force and Matter: The Conceptual Development of Nineteenth-Century Physics* (Cambridge: Cambridge University Press, 1982), p. 45, and Barri J. Gold, 'The Consolation of Physics: Tennyson's Thermodynamic Solution', *PMLA*, 117: 3 (May 2002), 449–64 (p. 450).

63 See Ted Underwood, 'How Did the Conservation of Energy become "The Highest Law in All Science?"' in *Repositioning Victorian Sciences: Shifting Centres in Nineteenth-Century Thinking*, ed. David Clifford and others (London: Anthem Press, 2006), pp. 119–30 (pp. 124–5).

64 See esp. Bruce Clarke, 'Allegories of Victorian Thermodynamics', *Configurations*, 4 (1996), 67–90, Gold, *Thermopoetics*, and Allen MacDuffie, 'Victorian Thermodynamics and the Novel: Problems and Prospects', *Literature Compass*, 8: 4 (2011), 206–13.

65 See Peggy Aldrich Kidwell, 'Prelude to Solar Energy: Pouillet, Herschel, Forbes and the Solar Constant', *Annals of Science*, 38 (1981), 457–76 (pp. 461, 475). John Herschel invented the actinometer to measure solar influx, but did not publish his results until the 1850s.

66 William Thomson, 'On the Mechanical Energies of the Solar System' (1854), *Transactions of the Royal Society of Edinburgh*, 21 (1857), 63–81 (p. 64).

67 Hermann von Helmholtz, 'Observations on the Sun's Store of Force', in *Early Solar Physics*, pp. 99–102 (p. 102).

68 William Thomson, 'On the Age of the Sun's Heat', *Macmillan's Magazine*, 5 (March 1862), 388–93 (p. 393). On the development of Thomson's ideas, see esp. Frank A. J. L. James, 'Thermodynamics and Sources of Solar Heat, 1846–1862', *British Journal for the History of Science*, 15: 2 (July 1982), 155–81.

69 Beer, 'The Death of the Sun', pp. 164–5, 170.

70 See Gold, 'The Consolation of Physics', p. 453, and MacDuffie, *Victorian Literature*, pp. 66–7.

71 Janssen, 'Notes on Recent Progress in Solar Physics', p. 168, Robert Ball, *The Story of the Sun* (London: Cassell, 1896), p. 270, and Charles William Siemens, 'On the Conservation of Solar Energy', *Proceedings of the Royal Society of London*, 33 (1882), 389–98 (p. 398).

72 Thomson, 'On the Mechanical Action of Radiant Heat or Light', p. 113. See Crosbie Smith, *The Science of Energy: A Cultural History of Energy Physics in Victorian Britain* (Chicago: The University of Chicago Press, 1998), p. 123.

73 See Augustin Mouchot, *La Chaleur Solaire et ses Applications Industrielles* (Paris: Gauthier-Villars, 1869), esp. pp. 96–100.

74 François Jarrige, '"Mettre le soleil en bouteille": les appareils de Mouchot et
    l'imaginaire solaire au début de la Troisième République', *Romantisme*, 150
    (2010), 85–96 (p. 93).

75 See Mouchot, *La Chaleur Solaire*, esp. pp. 229–33, and Jarrige, 'Mettre le soleil
    en bouteille', p. 97.

76 Charles Henry Pope, *Solar Heat: Its Practical Applications* (Boston, MA: by the
    author, 1903), pp. 36–7.

77 Perlin, *Let It Shine*, pp. 94–5.

78 Anon., 'The Place Where Light Dwelleth', p. 413.

79 See 'Stalk', *v.* 1d, and *v.*2, *OED Online*, accessed via <http://dictionary.oed.com>.

80 Anon., 'More Work for the Sun', p. 490. See also Anon., 'The Utilisation of Sun-
    Power', *Chambers's Journal of Popular Literature, Science and Arts*, Issue 807 (14
    June 1879), 736–77.

81 On solar ovens in the twentieth century, see Ethan Barnaby Kapstein, 'The Solar
    Cooker', *Technology and Culture*, 22: 1 (January 1981), 112–21.

82 See, for instance, John Ericsson, 'Solar Heat', *Nature*, 122: 2 (29 February 1872),
    344–7.

83 See John Ericsson, *Contributions to the Centennial Exhibition* (New York:
    The Nation Press, 1876), p. 559. For some speculation on Ericsson's solar
    technology, see L. C. Spencer, 'A Comprehensive Review of Small Solar-Powered
    Heat Engines: Part I', *Solar Energy*, 43: 4 (1989), 191–2 (p. 192). On the 'pencil'
    metaphor within discourses of light, see Alice Jenkins, *Space and the March
    of Mind: Literature and the Physical Sciences in Britain, 1815–1850* (Oxford:
    Oxford University Press, 2007), p. 178.

84 Nikola Tesla, meanwhile, dismissed the whole concept of the solar engine,
    particularly due to its low efficiency and the difficulty of storing the captured
    energy. See 'The Problem of Increasing Human Energy, with Special References
    to the Harnessing of the Sun's Energy', *Century* (June 1900), 175–216; extract
    in *The Tesla Papers*, ed. David Hatcher Childress (Kempton, IL: Adventures
    Unlimited Press, 2000), pp. 49–52 (p. 50).

85 See F. W. J. Hemmings, 'Émile Zola devant l'Exposition Universelle de 1878',
    Cahiers de l'Association internationale des études francaises, 24 (1972), 131–53,
    and Jarrige, 'Mettre le soleil en bouteille', p. 85.

86 See Julia Przybos, 'Zola's Utopias', in *The Cambridge Companion to Zola*, ed.
    Brian Nelson (Cambridge: Cambridge University Press, 2007), pp. 169–87 (p.
    183), and Eduardo A. Febles, 'The Anarchic Commune as World's Fair in Émile
    Zola's *Travail*', *Nineteenth-Century French Studies*, 36: 3&4 (Spring-Summer
    2008), 286–304 (esp. pp. 289–90).

87 Émile Zola, *Labor: A Novel*, trans. Anon. (New York: Harper, 1901), pp. 534–5.

88 In contrast, in Mark Twain's comic novel *The American Claimant* (1892), the eccentric American inventor Colonel Mulberry Sellers sees solar power as a route to individual profit, apparently having discovered how to apply the 'stupendous energies' of sunspots to the 'reorganizing of our climates', and seeks to purchase and transform Siberia himself. See Mark Twain, *The American Claimant* (New York: Charles L. Webster, 1892), p. 272, and James Rodger Fleming, *Fixing the Sky: The Checkered History of Weather and Climate Control* (New York: Columbia University Press, 2010), pp. 27–30.

89 Royal Astronomical Society Herschel MSS, W. 3/1. 4 ('Moon'), pp. 8, 9.

90 Washington Irving, *A History of New York*, 2 vols (New York: Inskeep & Bradford, 1809), I, 59.

91 See A. J. Meadows, *Science and Controversy: A Biography of Sir Norman Lockyer*, 2nd edn (London: Macmillan, 2008), p. 165.

92 See J. Norman Lockyer, *The Dawn of Astronomy: A Study of the Temple-Worship and Mythology of the Ancient Egyptians* (London: Cassell and Company, 1894), pp. 66, 104, 93, 23, and Wells, 'The Sun God and the Holy Stars', *Pall Mall Gazette*, 58 (24 February 1894), 3.

93 Ball, *The Story of the Sun*, pp. 61, 241–2; Charles Augustus Young, *The Sun* (New York: D. Appleton, 1881).

94 H. G. Wells, 'An Excursion to the Sun', *Pall Mall Gazette*, 58 (6 January 1894), 4.

95 H. G. Wells, *The Time Machine*, ed. Roger Luckhurst (Oxford: Oxford University Press, 2017), p. 78.

96 H. G. Wells, *The Discovery of the Future* (New York: B. W. Huebsch, 1913), p. 55.

97 I. W. Heysinger, *The Source and Mode of Solar Energy Throughout the Universe* (Philadelphia: J. B. Lippincott, 1895), pp. 9, 17, 343, 86–7 and 127.

98 H. G. Wells, 'Scientific Research as a Parlour Game', *The Saturday Review*, 79 (20 April 1895), 516.

99 Wells, *The War of the Worlds*, p. 52.

100 See David Ketterer, 'Introduction: "Flashes of Light"', in *Flashes of the Fantastic*, ed. Ketterer (Westport, CT: Praeger, 2004), pp. 1–6.

101 Percival Lowell, *Mars* (Boston: Houghton, Mifflin and Co., 1895), p. 209.

102 H. G. Wells, 'Intelligence on Mars', *The Saturday Review*, 81 (4 April 1896), 345–6 (p. 346). On Javelle's observations, see 'A Strange Light on Mars', *Nature*, 50 (1894), 319, repr. in *The War of the Worlds*, pp. 306–7.

103 See Francis Galton, 'Intelligible Signals between Neighbouring Worlds', *Fortnightly Review*, 60: 359 (November 1896), 657–64, Galton, 'Sun Signals for the Use of Travellers (Hand Heliostat)', *Proceedings of the Royal Geographical*

Society of London, 4: 1 (28 November 1859), 14–19, and Will Tattersdill, *Science, Fiction, and the Fin-de-Siècle Periodical Press* (Cambridge: Cambridge University Press, 2016), p. 30.

104 Aaron Worth, 'Imperial Transmissions: H. G. Wells, 1897–1901', *Victorian Studies*, 52 (2010), 65–89.

105 See Winston S. Churchill, *The Story of the Malakand Field Force: An Episode of Frontier War* (London: Thomas Nelson, 1916; repr. New York: Dover, 2010), pp. 225–6.

106 Rudyard Kipling, *Rudyard Kipling's Verse: Inclusive Edition, 1885–1918* (New York: Doubleday, Page & Co., 1920), pp. 13, 14.

107 Worth, 'Imperial Transmissions', pp. 70, 71; Alex Eisenstein, 'Very Early Wells: Origins of Some Major Physical Motifs in *The Time Machine* and *The War of the Worlds*', *Extrapolation*, 13 (1972), 119–26 (pp. 124–5); Wells, *Experiment in Autobiography: Discoveries and Conclusions of a Very Ordinary Brain (Since 1866)*, 2 vols (1934; London: Faber, 1984), I, 137.

108 M. A. Lacqui, 'Death-Rays and Moonshine: Is There a Menace?' *Conquest*, 5: 9 (1924), 382–3 (p. 382). See also Garrett P. Serviss's unauthorized sequel to *War*, *Edison's Conquest of Mars* (1898) (in which 'Edison' emulates Martian technology in designing a 'parabolic reflector' able to send 'destructive waves … like a beam of light, but invisible'), and Edgar Rice Burroughs's *A Princess of Mars* (1912) (in which the Martian Barsoom use the 'eighth ray' within light to propel their battle ships): Serviss, *Edison's Conquest of Mars* (1898), introduced by A. Langley Searles (Los Angeles: Carcosa House, 1947), p. 14, and Burroughs, *A Princess of Mars* (New York: Grosset & Dunlop, 1917), pp. 244, 245.

109 Herbert L. Sussman, *Victorians and the Machine: The Literary Response to Technology* (Cambridge, MA: Harvard University Press, 1968), p. 178.

110 On Promethean metaphors for industrial machinery, see Herbert L. Sussman, *Victorian Technology: Invention, Innovation, and the Rise of the Machine* (Santa Barbara, CA: ABC-CLIO, 2009), pp. 48–9.

111 See William Henry Fox Talbot, 'On the Production of Instantaneous Photographic Images', *Athenaeum*, 1258 (6 December 1851), 1286–7.

112 Frederick Soddy, *The Interpretation of Radium*, 2nd edn (London: John Murray, 1909), p. 239.

113 V. K. Zworykin and E. G. Ramberg, *Photo-Electricity and Its Application*, 2nd edn (New York: J. Wiley, 1949), pp. 176, 26.

114 See Willoughby Smith, 'The Action of Light on Selenium', *Journal of the Society of Telegraph Engineers*, 2 (1873), 31–3 (p. 32); Smith, *Selenium: Its Electrical Qualities, and the Effect of Light Thereon, Being a Paper Read before the Society*

of Telegraph Engineers, 28th November 1877 (London: Hayman Bros. and Lilly, 1877), pp. 9, 16; W. G. Adams and R. E. Day, 'The Action of Light on Selenium', *Philosophical Transactions*, 168 (1877), 313–49 (p. 342).

115 Maxwell to Peter Guthrie Tait, late April 1874, in *The Scientific Papers and Letters of James Clerk Maxwell*, ed. P. M. Harman, 2 vols (Cambridge: Cambridge University Press, 1990–2002), III, 67. See also Maxwell to George Gabriel Stokes, 31 October 1876, in *Scientific Papers*, III, 410–14.

116 See Alexander Graham Bell, 'The Photophone', *Science*, 1: 11 (11 September 1880), 130–4, and George M. Minchin, 'Experiments in Photoelectricity', *Proceedings of the Physical Society of London*, 11 (1890–92), 67–102 (p. 67).

117 M. Berthier, 'Electricity from Light', *The Photographic Times*, 25: 680 (28 September 1894), 209–10 (p. 209). On selenium's efficiency, see Zworykin and Ramberg, *Photo-Electricity*, p. 470.

118 Rollo Appleyard, 'Photo-Electric Cells', *The Telegraphic Journal and Electrical Review*, 28: 687 (23 January 1891), 124–6 (p. 125).

Chapter 5

1 Hugo Gernsback, 'The Utilization of the Sun's Energy', *The Electrical Experimenter*, 3: 11 (March 1916), 605–6, 662–3 (p. 663).

2 See Pope, *Solar Heat: Its Practical Applications*, pp. 13–14.

3 Graeme Macdonald, 'Improbability Drives: The Energy of SF', *Paradoxa*, 26 (2014), accessed via <http://strangehorizons.com/non-fiction/articles/improb ability-drives-the-energy-of-sf/>.

4 See, for instance, 'The Commercial Utilization of Solar Radiation and Wind Power', *Scientific American*, 105: 3 (21 January 1911), 65, 76–7.

5 See David E. Nye, *Consuming Power: A Social History of American Energies* (Cambridge, MA: The MIT Press, 1998), p. 157.

6 Quoted in Hugh Kenner, *A Homemade World: The American Modernist Writers* (1975; London: Marion Boyers, 1977), p. 5.

7 See Gernsback, 'Preface to the Second Edition', in *Ralph 124C 41+*, 2nd edn (New York: Frederick Fell, 1950), pp. 7–10 (p. 8).

8 See Fletcher Pratt, 'Foreword', in *Ralph 124C 41+*, 2nd edn, pp. 19–24 (pp. 21–2), and Gary Westfahl, *The Mechanics of Wonder: The Creation of the Idea of Science Fiction* (Liverpool: Liverpool University Press, 1998), p. 92.

9 Hugo Gernsback, *Ralph 124C 41+: A Romance of the Year 2660* (Lincoln: University of Nebraska Press, 2000), p. 99.

10 See Gary Westfahl, 'Evolution of Modern Science Fiction: The Textual History of Hugo Gernsback's *Ralph 124C 41+*', *Science Fiction Studies*, 23: 1 (March 1996), 37–82 (p. 43).

11 Winthrop Packard, 'Power from Sunlight', *The Technical World Magazine*, 11: 4 (June 1909), 356–60 (p. 357).

12 *The New York Herald* (19 October 1909), and *The World* (19 October 1909). See Dennis Bartels, 'George Cove's Solar Energy Device', *Material Culture Review*, 46 (Autumn 1997), 45–50 (esp. p. 46).

13 '"Sun-Harnessers" Alleged Cheats', *New York Times*, 10 August 1911, p. 7.

14 *Modern Electrics*, 4: 6 (September 1911), front cover, accessed via <http://ww w.philsp.com/homeville/SFI/t912.htm#A18580>.

15 Frank Shuman, 'Power from Sunshine: A Pioneer Solar Power Plant', *Scientific American*, 105: 14 (30 September 1911), 291–2, and Shuman, 'Feasibility of Utilizing Power from the Sun', *Scientific American*, 110: 9 (28 February 1914), 179. See also 'A Solar Engine', *The Diamond Fields Advertiser*, Issue 8971 (18 November 1907), p. 5.

16 See George Bronson Rea, ed., 'To Utilize the Tropical Sun', *The Far Eastern Review*, 11 (April 1908), 324–5, and Frank T. Kryza, *The Power of Light: The Epic Story of Man's Quest to Harness the Sun* (New York: McGraw-Hill, 2003), p. 2.

17 Anon., 'An Egyptian Solar Power Plant: Putting the Sun to Work', *Scientific American*, 107: 4 (25 January 1913), 88.

18 Shuman, 'Feasibility of Utilizing Power', p. 179, and see Perlin, *Let It Shine*, pp. 136–9.

19 Shuman, 'Feasibility of Utilizing Power', p. 179.

20 A. S. E. Ackermann, 'The Utilisation of Solar Energy', *The Electrical Experimenter*, 2: 2 (June 1914), 19; Anon., 'Energy from the Sun', *Scientific American*, 110: 21 (23 May 1914), 424.

21 Frank Shuman, 'Sun-power Plants Not Visionary', *Scientific American*, 110: 26 (27 June 1914), 519.

22 Perlin, *Let It Shine*, p. 142.

23 Waldemar Kaempffert, 'Harnessing Nature: Can the Free Energy of Space Be Utilized?' *Scientific American*, 108: 14 (5 April 1913), 308–9 (p. 308).

24 Charlotte Perkins Gilman's short story 'Beewise' (1913), for instance, imagines the utopian, 'absolutely self-supporting' binary towns Beewise and Herways, whose innovations include a 'solar engine' and hydroelectricity. See *Herland and Selected Short Fiction of Charlotte Perkins Gilman*, ed. Barbara H. Solomon (New York: Signet Classic, 1992), pp. 211–19 (pp. 218, 216); originally published in *The Forerunner*, 4: 7 (July 1913), 169–73.

25 Robert A. Heinlein, 'On the Writing of Speculative Fiction', in *Of Worlds Beyond: The Science of Science Fiction Writing*, ed. Lloyd Arthur Eshbach (1947; Chicago: Advent, 1964), pp. 13–19.

26 [Heinlein], 'Let There Be Light', *Super Science Stories* (May 1940), 34–45 (pp. 36, 35, 38).

27 E. Newton Harvey, *A History of Luminescence: From the Earliest Times Until 1900*, *Memoirs of the American Philosophical Society*, 44 (Philadelphia: American Philosophical Society, 1957), pp. 3–4, 243, vii, 418. See also Oliver Lodge, 'Radium and its Lessons', *The Nineteenth Century and After*, 54: 317 (July 1903), 76–85 (p. 85).

28 H. G. Wells, *The World Set Free: A Study of Mankind* (New York: E. P. Dutton, 1914), pp. 32–3.

29 See Imre Szeman, 'System Failure: Oil, Futurity, and the Anticipation of Disaster', *South Atlantic Quarterly*, 106: 4 (Fall 2007), 805–23 (p. 314), and Gerry Canavan, 'Retrofutures and Petrofutures: Oil, Scarcity, Limit', in *Oil Culture*, eds. Ross Barrett and Daniel Worden (Minneapolis, MN: University of Minnesota Press, 2014), pp. 331–49 (p. 337).

30 Hugo Gernsback, 'Cold Light', *The Electrical Experimenter*, 6: 7 (November 1918), 443. See also Hugo Gernsback, 'Cold Light', *Science and Invention*, 8: 9 (January 1921), 943.

31 Robert Heinlein Archive, Box 3, Document 3, OPUS004. Heinlein's early works often feature independent, intelligent female characters: see Helen Merrick, 'Gender in Science Fiction', in *The Cambridge Companion to Science Fiction*, eds. Edward James and Farah Mendlesohn (Cambridge: Cambridge University Press, 2003), pp. 241–52 (p. 245).

32 Heinlein Archive, Box 3, Document 3, OPUS004. Heinlein also toyed with 'The Power and the Glory' as a title, at a much later stage.

33 See Howard Bruce Franklin, *Robert A. Heinlein: America as Science Fiction* (Oxford: Oxford University Press, 1980), pp. 22–3, and Bill Patterson, 'A Study of "Let There Be Light"', *Heinlein Journal*, 4 (January 1999), 17–22 (pp. 19–20).

34 See William H. Patterson, Jr., *Robert A. Heinlein: In Dialogue with His Century: Volume 1 (1907–1948): Learning Curve* (New York: Tom Doherty Associates Books, 2010), pp. 239, 218.

35 See Upton Sinclair, *I, Governor of California and How I Ended Poverty: A True Story of the Future* (London: T. Werner Laurie, 1933), p. 10.

36 See, for instance, Robert Heinlein, 'Coventry', *Astounding Science Fiction*, 25: 5 (July 1940), 56–93 (p. 61).

37   Robert Heinlein, 'The Roads Must Roll', in *The Man Who Sold the Moon* (1953; London: Pan Books, 1955), pp. 51–89 (pp. 57, 58, 59, 75); originally published in *Astounding Science Fiction*, 25: 4 (June 1940), 2–22. See esp. Arthur C. Clarke, 'The Future of Transport', in *Profiles of the Future* (1962; London: Pan Books, 1964), pp. 30–43 (pp. 33–4).

38   Heinlein Archive, Box 3, Document 3, OPUS004.

39   See W. Olaf Stapledon, *Last and First Men: A Story of the Near and Far Future* (London: Methuen, 1930), p. 159, Stapledon, *Star Maker* (London: Methuen, 1937), pp. 158, 239, 234, and Freeman Dyson, 'Search for Artificial Stellar Sources of Infra-Red Radiation', *Science*, 131: 3414 (1960), 1667–8.

40   See *Isaac Asimov Presents the Best Science Fiction Firsts*, eds. Isaac Asimov and others (New York: Beaufort, 1984), p. 207, and Gary Westfahl, *Islands in the Sky: The Space Station Theme in Science Fiction Literature*, 2nd edn (1996; San Bernardino, CA: Borgo Press, 2009), p. 39.

41   Konstantin Tsiolkovsky, *Beyond the Planet Earth* (1920), trans. Kenneth Syers (Oxford: Pergamon Press, 1960); Otto Willi Gail, *The Stone from the Moon* (Baen Books, 2013). See also William J. Fanning, Jr., 'The Historical Death Ray and Science Fiction in the 1920s and 1930s', *Science Fiction Studies*, 37 (2010), 253–74 (pp. 260–1). On Oberth, see Goldberg, *The Mirror and Man*, pp. 228–9.

42   Murray Leinster, 'The Man Who Put Out the Sun', *Argosy*, 213: 1 (14 June 1930), 24–58, and see Everett F. Bleiler, *Science Fiction: The Early Years* (Kent, OH: The Kent State University Press, 1990), pp. 35–6.

43   Rosslyn D. Haynes, *From Faust to Strangelove: Representations of the Scientist in Western Literature* (Baltimore, MA: The Johns Hopkins University Press, 1994), p. 196.

44   *Murray Leinster: The Life and Works*, eds. Billee J. Stallings and Jo-an J. Evans (Jefferson, NC: McFarland, 2011), p. 3.

45   Murray Leinster, 'The Power Planet', *Amazing Stories: Scientific Fiction*, 6: 3 (June 1931), 198–217, 227 (p. 202).

46   Edmond Hamilton, 'Space Mirror', *Thrilling Wonder Stories: The Magazine of Prophetic Fiction*, 10: 1 (August 1937), 43–51 (p. 44).

47   Macdonald, 'Improbability Drives'.

48   See Adalberto Aguirre Jr., *Racial and Ethnic Diversity in America: A Reference Handbook* (Santa Barbara, CA: ABC-CLIO, 2003), p. 23.

49   S. K. Bernfeld, 'The Solar Menace', *Thrilling Wonder Stories*, 10: 1 (August 1937), 111–13 (pp. 111, 112, 113). Bernfeld's story is reminiscent of George Griffith's 'A Corner in Lightning', *Pearson's Magazine*, 5: 27 (March 1898), 264–71, in which a

power plant is built on the magnetic north pole to use its energy, but is destroyed in the process.

50 George Orwell, *Nineteen Eighty-Four: A Novel* (London: Secker & Warburg, 1949), p. 195 (Part 2, Section 9).

51 See Albert Einstein, 'A Heuristic Viewpoint on Generation and Modification of Light', in *Einstein in English: Papers on Physics*, trans. A. F. Kracklauer, 3 vols (Morrisville, NC: Lulu.com, 2010), I, 51–62, and Rachel Crossland, *Modernist Physics: Waves, Particles, and Relativities in the Writings of Virginia Woolf and D. H. Lawrence* (Oxford: Oxford University Press, 2018), pp. 23–4.

52 Cf. Graeme Macdonald, 'The Resources of Fiction', *Reviews in Cultural Theory*, 4: 2 (2013), 1–24 (p. 5) on 'narrative motion'.

53 Isaac Asimov, 'Runaround' (1942), in *I, Robot* (London: Ferndale, 1954), pp. 37–56. See also Clifford Simak's 'Masquerade', published a month before 'Reason' (and in the same magazine), which includes photocells on Mercury collecting solar energy and relaying it via the 'tight-beam principle' to substations orbiting other planets, similarly fantasizing an abundance of power (*Astounding Science Fiction*, 27: 1 (March 1941), 34–45 (pp. 36–7)).

54 Asimov, 'Reason' (1941), in *I, Robot*, pp. 57–76 (p. 59).

55 National Renewable Energy Laboratory 'Best Research-Cell Efficiencies' graph, accessed via <https://www.energy.gov/eere/solar/downloads/research-cell-efficiency-records>.

56 Isaac Asimov, *The Caves of Steel* (London: Science Fiction Book Club, 1956), p. 159.

57 Isaac Asimov, 'The Last Question' (1956), in *Space Opera: An Anthology of Way-Back-When Futures*, ed. Brian Aldiss (London: Weidenfeld and Nicolson, 1974), pp. 309–22 (p. 322).

58 Peter E. Glaser, 'Power from the Sun: Its Future', *Science*, 162: 3856 (22 November 1968), 857–61 (pp. 858, 859).

59 Peter E. Glaser, interviewed in Behrman, *Solar Energy*, p. 246.

60 Stacy V. Jones, 'A System for Solar Power Is Devised', Special to the *New York Times*, 29 December 1973, p. 31.

61 Krafft A. Ehricke, 'Space Light: Space Industrial Enhancement of the Solar Option', *Acta Astronautica*, 6 (1979), 1515–633 (p. 1518).

62 See Robert L. Heilbroner, *An Inquiry into The Human Prospect* (New York: Norton, 1974), pp. 51–3, and 'Fake Moon: Could China Really Light Up the Night Sky?' *BBC News Online*, 20 October 2018, accessed via <https://www.bbc.co.uk/news/world-asia-china-45910479>.

63 See esp. Kim Stanley Robinson's Mars Trilogy (1992–96).

64　Isaac Asimov, *The Gods Themselves* (New York: Doubleday, 1972), pp. 27, 56.

65　Many of the articles reprinted in Isaac Asimov, *The Beginning and the End* (New York: Doubleday, 1977) and *The Roving Mind* (1983; Oxford: Oxford University Press, 1987) address the issue of solar power, and particularly the SSPS concept. See also Asimov, 'Visit to the World's Fair of 2014', *New York Times*, 16 August 1964.

66　Isaac Asimov, 'Tighten Your Belt' (1974), in *The Beginning*, pp. 181–7 (p. 181).

67　Isaac Asimov, 'The Glorious Sun', in *The Beginning*, pp. 74–9 (p. 79); originally published as 'Solar Power: New Ways to Add to Our Energy Supply', *Mainliner Magazine*, December 1974. See also Asimov's primer for children, *How Did We Find Out about Solar Power?* (New York: Avon Books, 1981), pp. 54–8, and Isaac Asimov and Frederik Pohl, *Our Angry Earth: A Ticking Ecological Bomb* (New York: Tom Doherty Associates, 1991), pp. 218–21, 224–6.

68　Compare, for instance, David Dickson, *Alternative Technology and the Politics of Technical Change* (London: Fontana, 1974), p. 13.

69　Isaac Asimov, 'The Payoff in Space', in *The Roving Mind*, pp. 324–7 (p. 325).

70　Pat Stone, 'The Plowboy Interview: Isaac Asimov-Science, Technology … and Space!' (1980), in *Conversations with Isaac Asimov*, ed. Carl Freedman (Jackson: University Press of Mississippi, 2005), pp. 56–73 (p. 64).

71　Darrell Schweitzer, 'Isaac Asimov' (1981), in *Conversations*, pp. 74–84 (p. 75); Fred Jerome, 'Science and American Society' (1981), in *Conversations*, pp. 85–94 (p. 88).

72　Harry Harrison, *Skyfall* (1976; London: Panther Books, 1985), p. 7.

73　See 'Cells Convert Power from Sun', Special to the *New York Times*, 15 May 1973, p. 14.

74　See Nina Möllers, 'Electrifying the World: Representations of Energy and Modern Life at World's Fairs, 1893–1982', in *Past and Present Energy Societies: How Energy Connects Politics, Technologies and Cultures*, eds. Nina Möllers and Karin Zachmann (Bielefeld: Transcript, 2012), pp. 45–78 (p. 75).

75　See Spencer R. Weart, *The Discovery of Global Warming*, revised and expanded edn (Cambridge, MA: Harvard University Press, 2008), pp. 38, 87, 99, 102–3.

76　Arthur Herzog, *Heat* (1977; London: Pan Books, 1978), p. 9.

77　*Policy Implications of Global Warming: Mitigation, Adaptation, and the Science Base* (Washington, DC: National Academy Press, 1992), p. 448, citing J. T. Early, 'Space-Based Solar Shield to Offset Greenhouse Effect', *Journal of the British Interplanetary Society*, 42 (December 1989), 567–9.

78　Joan-Pau Sánchez and Colin R. McInnes, 'Optimal Sunshade Configurations for Space-Based Geoengineering near the Sun-Earth L_1 Point', *PLoS ONE*, 10: 8 (August 2015), 1–25 (p. 22).

79 Marcus Sedgwick and others, *Dark Satanic Mills* (London: Walker Books, 2013), pp. 18–19.

80 On 'dazzled reason', see Michel Foucault, *Madness and Civilization: A History of Insanity in the Age of Reason*, trans. Richard Howard (1967; London: Routledge, 1995), p. 108.

Chapter 6

1 David Clay Large, *Nazi Games: The Olympics of 1936* (New York: Norton, 2007), pp. 3, 9.

2 Robert K. Barney and Anthony Th. Bijerk, 'The Genesis of Sacred Fire in Olympic Ceremony: A New Interpretation', *Journal of Olympic History*, 13: 2 (May/June 2005), 6–27 (pp. 8, 18n70).

3 Nye, *Narratives and Spaces*, p. 88, and Lakoff and Johnson, *Metaphors We Live By*, p. 97.

4 D. H. Lawrence, 'A Propos of "Lady's Chatterley's Lover"' (1930), in *Lady Chatterley's Lover*, ed. Michael Squires (Cambridge: Cambridge University Press, 1993), pp. 303–35 (p. 331). Lawrence wrote several stories centred around pseudo-religious solar experience, including 'Sun' and 'The woman who rode away', in *The Woman Who Rode Away and Other Stories* (London: Martin Secker, 1928; repr. 1931), pp. 33–54, and 57–102. On the possible influence of sunlight therapy on 'Sun', see Kirsty Martin, 'Modernism and the Medicalization of Sunlight: D. H. Lawrence, Katherine Mansfield and the Sun Cure', *Modernism/modernity*, 23: 2 (2016), 423–41.

5 A. S. Eddington, 'On the Radiative Equilibrium of the Stars', *Monthly Notices of the Royal Astronomical Society*, 77 (November 1916), 16–35 (pp. 23, 20), and Eddington, 'The Internal Constitution of the Stars', *The Scientific Monthly*, 11: 4 (October 1920), 297–303 (p. 299). See also 'A Vision of Energy', *The Times*, Issue 45547 (24 June 1930), 13.

6 Wells, *The World Set Free*, pp. 14–15, 40.

7 Robert Frost, 'At Woodward's Gardens' (1936), in *Collected Poems, Prose, & Plays*, eds. Richard Poirier and Mark Richardson (New York: Library of America, 1995), p. 266 (l. 2).

8 See *The Notebooks of Robert Frost*, ed. Robert Faggen (Cambridge, MA: Harvard University Press, 2007), p. 656, and 'The Future of Man', in *The Collected Prose of Robert Frost*, ed. Mark Richardson (Cambridge, MA: Belknap Press of Harvard University Press, 2007), p. 209.

9 For other examples, see Elizabeth DeLoughrey, 'Heliotropes: Solar Ecologies and Pacific Radiations', in *Postcolonial Ecologies: Literatures of the Environment*, eds. DeLoughrey and George B. Handley (Oxford: Oxford University Press, 2011), pp. 235–53.

10 H. S. Truman, White House statement, Washington DC (6 August 1945); Robert Jungk, *Brighter Than a Thousand Suns: The Moral and Political History of the Atomic Scientists* (1956), trans. James Cleugh (London: Gollancz, 1958), p. 198.

11 See Randall Jarrell, '1945: The Death of the Gods' (1948), l. 5, in *The Complete Poems* (London: Faber, 1971), p. 183.

12 David Dietz, *Atomic Energy in the Coming Era* (New York: Dodd, Mead & Company, 1945), pp. 16, 19.

13 Ray Bradbury, 'The Golden Apples of the Sun', in *The Golden Apples of the Sun* (New York: Doubleday, 1953), pp. 243–50 (p. 244). See Arthur C. Clarke's discussion of the story in 'You Can't Get There from Here', in *Profiles of the Future* (1962; London: Pan Books, 1964), pp. 98–110 (pp. 107–8).

14 Charles Morgan, *The Burning Glass*, 2nd edn (London: Macmillan, 1955; repr. 1961), pp. 65, xvi, and 72.

15 On cloud-seeding, see Fleming, *Fixing the Sky*, esp. pp. 186, 194.

16 Despite this, Charles A. Carpenter excludes Morgan's play from his study of nuclear drama: see *Dramatists and the Bomb: American and British Playwrights Confront the Nuclear Age, 1945–1964* (Westport, CN: Greenwood Press, 1999), p. 7.

17 See Charles Thorpe, *Oppenheimer: The Tragic Intellect* (Chicago: Chicago University Press, 2006), Chapter 7.

18 See Lorna Arnold, *A Very Special Relationship: British Atomic Weapon Trials in Australia* (London: HMSO, 1987), pp. 6–7.

19 See Peter Hennessy, *The Secret State: Whitehall and The Cold War*, revised edn (London, Penguin, 2003), pp. 32–3, 90, and Simone Turchetti, 'Atomic Secrets and Governmental Lies: Nuclear Science, Politics and Security in the Pontecorvo Case', *The British Journal for the History of Science*, 36: 4 (December 2003), 389–415.

20 Chapman Pincher, 'How Much Would Pontecorvo Be Worth to Russia?' *Daily Express*, 24 October 1950, p. 4.

21 Harry S. Truman, 'Special Message to Congress Presenting a 21-Point Program for the Reconversion Period', 6 September 1945, *Public Papers of the Presidents of the United States, Harry S. Truman, 1945* (Washington, DC: Office of the Federal Register, 1961; repr. Ann Arbor, MI: University of Michigan Library, 2005), p. 262, accessed via <http://name.udml.umich.edu/4728442.1945.001>.

22 See D. H. Dowling, 'The Atomic Scientist: Machine or Moralist?' *Science-Fiction Studies*, 13: 2 (July 1986), 139–47.

23 See Kirsten Shepherd-Barr, *Science on Stage: From Doctor Faustus to Copenhagen* (Princeton: Princeton University Press, 2006), esp. pp. 20, 63.

24 Robert S. Oppenheimer, 'Physics in the Contemporary World', *Bulletin of the Atomic Scientists*, 4: 3 (March 1948), 65–8, 85–6 (pp. 65, 67), and P. W. Bridgman, 'Scientists and Social Responsibility', *Bulletin of the Atomic Scientists*, 4: 3 (March 1948), 69–72 (p. 70).

25 Morgan to Sir Carleton Allen, 18 March 1954, in *Selected Letters of Charles Morgan*, ed. Eiluned Lewis (London: Macmillan, 1967), p. 208.

26 Haynes, *From Faust to Strangelove*, p. 303.

27 Gerard Weales, 'The Devil's Glass', *Educational Theatre Journal*, 10: 4 (December 1958), 311–15 (p. 314), and Brooks Atkinson, 'Burning Glass: New Melodrama Contains Provocative Idea', *New York Times*, 14 March 1954, p. xi.

28 See Wallace Robbins, 'Review of Charles Morgan, *The Burning Glass*', *Bulletin of Atomic Scientists*, 12: 2 (February 1956), 60–1 (p. 61).

29 Oppenheimer, 'Physics in the Contemporary World', p. 66.

30 Kevin McCarron, *William Golding*, 2nd edn (Tavistock: Northcote House, 2006), p. 3; William Golding, *Lord of the Flies* (1954; London: Faber, 1962), p. 105.

31 John Carey, 'William Golding Talks to John Carey', in *William Golding: The Man and His Books*, ed. Carey (London: Faber, 1986), pp. 171–89 (p. 183).

32 J. P. Stern, *On Realism* (London: Routledge & Kegan Paul, 1973), pp. 20–1.

33 R. M. Ballantyne, *The Coral Island: A Tale of the South Seas* (London: Collins, 1906), p. 32.

34 See W. E. Davis, 'Mr. Golding's Optical Delusion', *English Language Notes*, 3: 2 (1965), 125–6.

35 Rohitash Thapliyal and Shakuntala Kunwar, 'Ecocritical Reading of William Golding's *Lord of the Flies*', *The IUP Journal of English Studies*, 6: 1 (2011), 85–90.

36 Clarke, 'Voices from the Sky', in *Profiles*, pp. 178–87 (p. 179). 'Extraterrestrial Relays', *Wireless World* (October 1946), is the original, groundbreaking essay.

37 Arthur C. Clarke, 'The Sun', in *The Challenge of the Spaceship: Previews of Tomorrow's World* (London: Frederick Muller, 1960), pp. 98–107 (pp. 104, 102); Arthur C. Clarke, 'The Radio Universe', in *The Challenge*, pp. 194–200 (pp. 195–6).

38 Arthur C. Clarke, 'Ages of Plenty', in *Profiles*, pp. 136–49 (p. 139); Neil McAleer, *Visionary: The Odyssey of Sir Arthur C. Clarke* (London: Victor Gollancz, 1992), after p. 370.

39 Clarke, 'The Sun', p. 100, with similar arguments made elsewhere.

40 Clarke, 'Ages of Plenty', p. 139.

41 E. N. Parker, 'Dynamics of the Interplanetary Gas and Magnetic Fields', *The Astrophysical Journal*, 128 (1958), 664–76 (p. 672), and Kenneth R. Lang, *The Cambridge Encyclopaedia of the Sun* (Cambridge: Cambridge University Press, 2001), p. 148. Clarke was perhaps also inspired by W. Olaf Stapledon's *Last and First Men* (1930), in which the last humans devise 'electro-magnetic "wave-systems" … capable of sailing forward upon the hurricane of solar radiation' (*Last and First Men*, p. 342).

42 See <https://www.nasa.gov/mission_pages/tdm/solarsail/index.html>. On the meeting with Gagarin, see McAleer, *Visionary*, p. 173.

43 On Clarke's reading, see McAleer, *Visionary*, pp. 20–1.

44 Arthur C. Clarke, 'The Stroke of the Sun', *Galaxy Magazine*, 16: 5 (September 1958), 71–7 (pp. 74, 73). It was renamed 'A Slight Case of Sunstroke' when republished in *Tales of Ten Worlds* (New York: Harcourt, Brace, & World, 1962), which also reprints another parodic death ray story 'Let There Be Light' (1957). Compare a story by the passive solar engineer Steve Baer (b.1938), in *Sunspots: Collected Facts and Solar Fiction* (Albuquerque, NM: Zomeworks, 1977), pp. 5–7, in which large mirrors start a riot. On Baer, see Alexis Madrigal, *Powering the Dream: The History and Promise of Green Technology* (Cambridge, MA: Da Capo Press, 2011), pp. 181–2, 186–7, and Andrew G. Kirk, *Counterculture Green: The Whole Earth Catalog and American Environmentalism* (Lawrence: University Press of Kansas, 2007), pp. 150–3.

45 See Arthur C. Clarke, *Tales from the White Hart* (New York: Ballatine Books, 1957).

46 Adam Roberts, *The History of Science Fiction* (Basingstoke: Palgrave Macmillan, 2006), pp. 212–13. Physicists A. A. Mill and R. Clift cite Clarke's tale when calculating that Archimedes's mirrors would have ignited not the Roman ships, but the passengers. See 'Reflections up the Mirrors of Archimedes, with a Consideration of the Geometry and Intensity of Sunlight Reflected from Plane Mirrors', *European Journal of Physics*, 13 (1992), 268–79 (pp. 278–9).

47 George R. Harrison, 'The Control of Energy', *The Atlantic Monthly*, 196: 3 (September 1955), 38–43 (p. 42). See Clarke, 'Ages of Plenty', p. 139, and 'The Sun', p. 100.

48 See D. M. Chapin, C. S. Fuller and G. L. Pearson, 'A New Silicon *p-n* Junction Photocell for Converting Solar Radiation into Electrical Power', *Journal of Applied Physics*, 25 (May 1954), 676–7, Chapin, Fuller and Pearson, 'The Bell Solar Battery', *Bell Laboratories Record*, 33 (July 1955), 241–6, and Perlin, *Let It Shine*, pp. 310–14.

49 Anon., 'Vast Power of the Sun Is Tapped by Battery Using Sand Ingredient', *New York Times*, 26 April 1954, p. 1.

50 Bell Labs, 'Bell System Solar Battery Converts Sun's Rays into Electricity!', advertisement in *Life*, 41: 15 (8 October 1956), p. 3.

51 See Harvey Strum, 'The Association for Applied Solar Energy/Solar Energy Society, 1954–1970', *Technology and Culture*, 26: 3 (July 1985), 571–8 (pp. 573–6).

52 See 'Original Vanguard Still Transmits after 4 Years', Special to the *New York Times*, 17 March 1962, p. 3.

53 See John Perlin, *From Space to Earth: The Story of Solar Electricity* (Ann Arbor, MI: Aatec Publications, 1999), pp. 41–2, 45, 50. On advances to solar cell design for space see, for example, John Noble Wilford, 'A New Solar Cell Resists Radiation', *New York Times*, 7 August 1966, p. 40.

54 Denis Hayes, 'Solar Possibilities', *Energy: The International Journal*, 4 (1979), 761–8 (p. 762).

55 Wallace Stevens, 'The Planet on the Table', in *The Collected Poems of Wallace Stevens* (New York: Alfred A. Knopf, 1954), p. 532. See John Tyndall, 'Scientific Use of the Imagination', in *Essays on the Use and Limit of the Imagination in Science* (London: Longmans, Green, and Co, 1870), pp. 13–51 (p. 47).

56 See esp. Jonathan Bate, *The Song of the Earth* (London: Picador, 2000; repr. 2001), pp. 282–3.

57 Philip Larkin, 'Solar', in *High Windows* (London: Faber, 1974), p. 33 (ll. 6, 21, 18–19).

58 Thom Gunn, 'Sunlight', in *Collected Poems* (London: Faber, 1994), p. 223 (ll. 22, 14, 15).

59 Otto Binder and Al Plastino, *Superman*, 146 (July 1961), p. 11.

60 Roy Thomas and Werner Roth, 'Call Him … Cyclops', *X-Men*, 43 (10 April 1968), repr. in *Giant-Size X-Men*, 1 (May 1975), n.p.

61 *Doctor Who: The Dalek Invasion of Earth* (BBC, November–December 1964), writer Terry Nation, dir. Richard Martin; *Doctor Who: The Daleks* (BBC, December 1963–February 1964), writer Terry Nation, dir. Christopher Barry and Richard Martin.

62 *Doctor Who: The Enemy of the World* (BBC, December 1967–January 1968), writer David Whitaker, dir. Barry Letts, Episode 2.

63 *Doctor Who: The Seeds of Death* (BBC, January–March 1969), writers Brian Hayles and Terence Dicks, dir. Michael Ferguson.

64 *Doctor Who: The Ark in Space* (BBC, January–February 1975), writer Robert Holmes, dir. Rodney Bennett.

65 *The Man with the Golden Gun*, dir. Guy Hamilton (Eon Productions, 1974); James Chapman, *Licence to Thrill: A Cultural History of the James Bond Films*, 2nd edn (London: Tauris, 2007), p. 145.

66 On sexualized technology in Bond, see Martin Willis, 'Hard-wear: The Millennium, Technology, and Brosnan's Bond', in *The James Bond Phenomenon: A Critical Reader*, ed. Christoph Lindner, 2nd edn (Manchester: Manchester University Press, 2003), pp. 169–83.

67 *Die Another Day*, dir. Lee Tamahori (MGM, 2002), DVD interview with Neal Purvis and Robert Wade.

68 *Megamind*, dir. Tom McGrath (DreamWorks, 2010); *Star Wars: Episode VII – The Force Awakens*, dir. J. J. Abrams (Lucasfilm/ Bad Robot Productions, 2015).

Chapter 7

1 'Who, what, why: How does a skyscraper melt a car?', *BBC News Online*, 3 September 2013, accessed via <https://www.bbc.co.uk/news/magazine-23944679>; Jon Henley, 'From the Walkie Talkie to the Death Ray Hotel: Buildings turn up the heat', *The Guardian*, 3 September 2013, accessed via <https://www.theguardian.com/artanddesign/shortcuts/2013/sep/03/walkie-talkie-death-ray-buildings-heat>; Oliver Wainwright, 'The Walkie-Talkie skyscraper, and the City's burning passion for glass', *The Guardian*, 3 September 2013, accessed via <https://www.theguardian.com/commentisfree/2013/sep/03/walkie-talkie-skyscraper>.

2 Jon Swaine, 'Guests burned by "death ray" from Las Vegas hotel', *The Daily Telegraph*, 29 September 2010, accessed via <https://www.telegraph.co.uk/news/worldnews/northamerica/usa/8031620/Guests-burned-by-death-ray-from-Las-Vegas-hotel.html>.

3 Oliver Wainwright, 'Walkie Talkie architect "didn't realise it was going to be so hot"', *The Guardian*, 6 September 2013, accessed via <https://www.theguardian.com/artanddesign/2013/sep/06/walkie-talkie-architect-predicted-reflection-sun-rays>.

4 *Stewart Lee's Comedy Vehicle*, Series 3: Episode 5, BBC2, first broadcast Saturday, 29 March 2014; writer Stewart Lee, dir. Tim Kirby.

5 David E. Nye, *American Technological Sublime* (Cambridge, MA: MIT Press, 1994), p. 106.

6 On this term, see esp. Jan Zalasiewicz and others, eds., *The Anthropocene as a Geological Time Unit: A Guide to the Scientific Evidence and Current Debate* (Cambridge: Cambridge University Press, 2019).

7   See Timothy Clark, 'Scale', in *Telemorphosis: Theory in the Era of Climate Change, Vol. 1*, ed. Tom Cohen (Michigan: Open Humanities Press, 2012), accessed via <https://quod.lib.umich.edu/o/ohp/10539563.0001.001/1:8/--telemorphosis-the ory-in-the-era-of-climate-change-vol-1?rgn=div1;view=fulltext>.

8   Spencer R. Weart, *The Discovery of Global Warming*, revised and expanded edn (Cambridge, MA: Harvard University Press, 2008), pp. 23, 112, 166.

9   V. Masson-Delmotte and others, eds., *Global Warming of 1.5°C: An IPCC Special Report* (Geneva: World Meteorological Organization, 2018), accessed via <https://www.ipcc.ch/sr15/>.

10 See Rowena Mason, 'Cameron at centre of "get rid of all the green crap" storm', *The Guardian*, 21 November 2013, accessed via <https://www.theguardian.com/e nvironment/2013/nov/21/david-cameron-green-crap-comments-storm>, Donald J. Trump, *Crippled America: How to Make America Great Again* (London: Simon & Schuster, 2015), p. 65, and 'Trump proposes solar panel wall for Mexico border', *BBC News Online*, 22 June 2017, accessed via <https://www.bbc.co.uk/news/av/worl d-us-canada-40365031/trump-proposes-solar-panel-wall-for-mexican-border>.

11 See esp. John Perlin, *From Space to Earth: The Story of Solar Electricity* (Ann Arbor, MI: Aatec Publications, 1999), pp. 172–8.

12 See Chris Goodall, *The Switch* (London: Profile Books, 2016), pp. 80, 82, 24–6, and Jon Major, 'The Search for Silicon's Successor', *Physics World*, 21 November 2018, accessed via <https://physicsworld.com/a/the-search-for-silicons-successor/>.

13 Thereby conforming somewhat to the post-peak oil genre identified by Gerry Canavan: see 'Retrofutures and Petrofutures: Oil, Scarcity, Limit', in *Oil Culture*, eds. Ross Barrett and Daniel Worden (Minneapolis, MN: University of Minnesota Press, 2014), pp. 331–49 (p. 332).

14 David Dickson, *Alternative Technology and the Politics of Technical Change* (London: Fontana, 1974), p. 13.

15 J. G. Ballard, 'The Ultimate City', in *Low-Flying Aircraft and Other Stories* (London: Jonathan Cape, 1976), pp. 7–87 (p. 14).

16 See esp. Andrzej Gasiorek, *J. G. Ballard* (Manchester: Manchester University Press, 2005), pp. 22–3, and Umberto Rossi, 'Shakespearean Reincarnations: An Intertextual Reading of J. G. Ballard's *The Ultimate City*', *Journal of the Fantastic in the Arts*, 20: 3 (2009), 363–84.

17 See Anon., 'Sun Power in the Pyrenees', *Time*, 95: 20 (18 May 1970), Félix Trombe and Albert Le Phat Vinh, 'Thousand kW Solar Furnace, Built by the National Center of Scientific Research, in Odeillo (France)', *Solar Energy*, 15 (1973), 57–61, and Dickson, *Alternative Technology*, p. 115.

18 Daniel Behrman, *Solar Energy: The Awakening Science* (London: Routledge & Kegan Paul, 1979), p. 29.

19 Peter Harper, 'Alternative Technology Revisited', *Undercurrents*, 50 (February 1982), 33–5 (p. 33).

20 See Andrew G. Kirk, *Counterculture Green: The Whole Earth Catalog and American Environmentalism* (Lawrence: University Press of Kansas, 2007), esp. pp. 150–3.

21 By contrast, the Street Farm Collective, one of the UK's other eco-communities, was based in southeast London and later, Bristol. See Stephen E. Hunt, *The Revolutionary Urbanism of Street Farm: Eco-anarchism, Architecture and Alternative Technology in the 1970s* (Bristol: Tangent Books, 2014), p. 22.

22 Duff Hart Davis, 'Design for No-Waste Living', *The Sunday Telegraph*, 27 October 1974, p. 8.

23 Paula Davies, 'Opting for the simpler life', *The Daily Telegraph*, 4 August 1982, p. 11. On similarly negative attitudes to renewables enthusiasts in the American media, see Finis Dunaway, *Seeing Green: The Use and Abuse of American Environmental Images* (Chicago: The University of Chicago Press, 2015), p. 165. On 1970s Britain as a place of significant research interest in solar, see Behrman, *Solar Energy*, pp. 261, 284.

24 On campaigns against economic growth, see Hunt, *The Revolutionary Urbanism*, p. 27.

25 In *Hello America* (1981), Ballard revisits the premise of a post-industrial West left abandoned after the exhaustion of fossil fuels. In contrast to 'The Ultimate City', it ends somewhat triumphantly, with the glimpse of a positive future heralded by a fleet of 'sunlight fliers' which use solar energy to produce a cushion of warm air upon which they glide (p. 167). See *Hello America* (1981; New York: Carroll & Graf, 1988), pp. 167, 220–4.

26 Naomi Klein, *This Changes Everything: Capitalism vs. the Climate* (London: Allen Lane, 2014), p. 279.

27 Peter Hitchcock, 'Oil in an American Imaginary', *New Formations*, 69: 4 (2010), 81–97 (p. 81).

28 We can contrast Ballard's story, therefore, with Ernest Callenbach's techno-utopian mock-travel novel *Ecotopia* published a year earlier which, in describing the renewable technologies of near-future California, resembles Gernsback's *Ralph 124C 41+* in its narrative stasis. See Ernest Callenbach, *Ecotopia: A Novel about Ecology, People and Politics in 1999* (1975; London: Pluto Press, 1978), pp. 103–5, and Kirk, *Counterculture Green*, pp. 156–8.

29 Norman Rush, *Mating* (London: Vintage, 1991), pp. 83, 85, 86.

30 See S. Ekema Agbaw and Karson L. Kiesinger, 'The Reincarnation of Kurtz in Norman Rush's *Mating*', *Conradiana*, 32: 1 (2000), 47–56.

31 See esp. Jimmy Carter, 'Address to the Nation on Energy' (18 April 1977), 'Solar Energy Remarks Announcing Administration Proposals' (20 June 1979), and 'Address to the Nation on Energy and National Goals: "The Malaise Speech"' (15 July 1979), accessed via *The American Presidency Project*, eds. John T. Woolley and Gerhard Peters <https://www.presidency.ucsb.edu/documents/>.

32 Reported in Martin Tolchin, 'Carter Welcomes Solar Power', Special to *The New York Times*, 21 June 1979, p. D1.

33 Dunaway, *Seeing Green*, pp. 171, 173–4, 179.

34 R. J. Martin-Palma and A. Lakhtakia, 'Engineered biomimicry for harvesting solar energy: A bird's eye view', *International Journal of Smart and Nano Materials*, 4: 2 (2013), 83–90.

35 Stephen Poliakoff, *Blinded by the Sun & Sweet Panic* (Portsmouth: Methuen, 1996), p. 31 (Act I, Scene IV).

36 Chris Johnston, 'Honest scientist (batteries not included): An Interview with Stephen Poliakoff', *The Times Higher Education Supplement*, Issue 1259 (20 December 1996), p. 16. On 'cold fusion', see Adil E. Shamoo and David B. Resnik, *Responsible Conduct of Research*, second edn (Oxford: Oxford University Press, 2009), p. 145.

37 Steve Jones, 'Poliakoff's dramatic appliance of science', *The Daily Telegraph*, 31 August 1996, accessed via <https://www.telegraph.co.uk/culture/4703799/Polia koffs-dramatic-appliance-of-science.html>.

38 Ian McEwan, *Solar* (2010; London: Vintage, 2011), p. 25.

39 'halo, n.' and 'energy, n.', *OED Online*, accessed via <http://dictionary.oed.com>.

40 Keith Barnham, *The Burning Answer: A User's Guide to the Solar Revolution* (London: Weidenfeld & Nicolson, 2014), p. 219.

41 See, for instance, George Monbiot, *Heat* (2006; London: Penguin, 2007), p. 125.

42 Ian McEwan, *Saturday* (2005; London: Vintage, 2006); Greg Garrard, '*Solar*: Apocalypse not', in *Ian McEwan: Contemporary Critical Perspectives*, ed. Sebastian Groes, second edition (London: Bloomsbury, 2013), pp. 123–36 (p. 126).

43 On nanostructuring, see Martin-Palma and Lakhtakia, 'Engineered biomimicry', p. 85.

44 See esp. Garrard, '*Solar*: Apocalypse not', p. 123.

45 See John Milton, *Paradise Lost*, ed. Alastair Fowler, second edn (London: Longman, 2006), p. 169 (III, 51).

46 Such as the 'Cash for Ash'/ Renewable Heat Incentive scandal which came to light in Northern Ireland in 2016.

47 Adam Trexler, *Anthropocene Fictions: The Novel in a Time of Climate Change* (Charlottesville: University of Virginia Press, 2015), pp. 34, 65.

48 Some grand solar schemes in the United States have encountered planning
    problems, ostensibly because of their potential damage to the natural
    environment/view. See Todd Woody, 'Desert Vistas vs. Solar Power', *New York
    Times*, 22 December 2009, B1. The risk-return profile of solar is much more
    attractive to institutional investors now, compared to 2010. See Bruce Usher,
    Renewable Energy (New York: Columbia University Press, 2019), p. 108, and also
    Stewart Brand, *Whole Earth Discipline* (2009; London: Atlantic Books, 2010),
    p. 102.

49 Here I adopt terms from Justin Good and Timothy Morton, respectively. See
    Good, 'The Aesthetics of Wind Energy', *Human Ecology Review*, 13: 1 (2006),
    76–89 (p. 80), and Morton, *Hyperobjects: Philosophy and Ecology after the End
    of the World* (Minneapolis: University of Minnesota Press, 2013), p. 105. On
    US solar farms, see esp. Alexis Madrigal, *Powering the Dream: The History and
    Promise of Green Technology* (Cambridge, MA: Da Capo Press, 2011), pp. 274–88.

50 *Blade Runner 2049* (Alcon Entertainment/ Columbia Pictures, 2017), dir. Denis
    Villeneuve; writers Hampton Fancher and Michael Green.

51 <http://www.johngerrard.net/solar-reserve.html>.

52 Kenneth Brower, 'The Danger of Cosmic Genius', *The Atlantic*, 27 October 2010,
    accessed via <https://www.theatlantic.com/magazine/archive/2010/12/the-da
    nger-of-cosmic-genius/308306/>.

53 Matthew Schneider-Mayerson, 'The Influence of Climate Fiction: An Empirical
    Survey of Readers', *Environmental Humanities*, 10: 2 (November 2018), 473–500
    (p. 495).

54 Jeremy Leggett, *Half Gone: Oil, Gas, Hot Air and the Global Energy Crisis* (2005;
    London: Portobello Books, 2006), pp. 240, 241. The chemist Paul Crutzen,
    famous for identifying the Anthropocene, is a proponent of injecting sulphate
    aerosols: see Klein, *This Changes Everything*, pp. 261–2.

55 Katherine Ellsworth-Krebs and Louise Reid, 'Conceptualising energy
    prosumption: Exploring energy production, consumption and microgeneration
    in Scotland, UK', *Environment and Planning A*, 48: 10 (2016), 1988–2005.

56 See Aimie Hope, Thomas Roberts, and Ian Walker, 'Consumer engagement in
    low-carbon home energy in the United Kingdom: Implications for future energy
    system decentralization', *Energy Research & Social Science*, 44 (2018), 362–70.

57 Gill Seyfang, Jung Jin Park, and Adrian Smith, 'A thousand flowers blooming? An
    examination of community energy in the UK', *Energy Policy*, 61 (2013), 977–89
    (p. 988).

58 Derek Mahon, 'Homage to Gaia: Its Radiant Energies', in *Life on Earth* (Oldcastle,
    County Meath, Ireland: The Gallery Press, 2008), pp. 44–5.

59   See Sam Solnick, *Poetry and the Anthropocene: Ecology, Biology and Technology in Contemporary British and Irish Poetry* (Abingdon, Oxon: Routledge, 2016), p. 130.

60   See <https://littlesun.com/about>.

61   On social sculpture, see Andrea Gyorody, 'The medium and the message: art and politics in the work of Joseph Beuys', *The Sixties*, 7: 2 (2014), 117–37 (p. 123).

62   Andrew Dincher, 'Foreword: On the Origins of Solarpunk', in *Sunvault: Stories of Solarpunk and Eco-Speculation*, eds. Phoebe Wagner and Brontë Christopher Wieland (Nashville, TN: Upper Rubber Boot, 2017), pp. 7–8 (p. 7).

63   Wagner and Wieland, 'Editors' Note', in *Sunvault*, pp. 9–10 (p. 9).

64   Sara Norja, 'Sunharvest Triptych', in *Sunvault*, pp. 218–20 (p. 219; ll. 38–40).

65   David A. Mathisen, '2079: A century of technical and socio-political evolution', *Impact of Science on Society*, 29: 1 (January–March 1979), 83–91 (p. 83).

66   Barnham, *The Burning Answer*, pp. 237, 142, 313–15, 273–81, 302–10, 358.

67   Ursula K. Heise, *Imagining Extinction: The Cultural Meanings of Endangered Species* (Chicago: The University of Chicago Press, 2016), p. 215.

68   Jonathan Porritt, *The World We Made: Alex McKay's Story From 2050* (London: Phaidon Press, 2013), p. 37.

69   Usher, *Renewable Energy*, pp. 13, 97–8.

70   Bill McKibben, 'To stop global catastrophe, we must believe in humans again', *The Guardian*, 23 April 2019, accessed via <https://www.theguardian.com/commentisfree/2019/apr/23/stop-global-catastrophe-believe-humans-again-geoengineering>

71   Usher, *Renewable Energy*, p. 128.

Selected bibliography

Primary sources

Adams, W. G. and R. E. Day, 'The Action of Light on Selenium', *Philosophical Transactions* (hereafter *PT*), 168 (1877), 313–49.

Allott, Miriam, ed., *The Poems of John Keats* (London: Longman, 1970).

Anon, 'An Account from Paris Concerning a great Metallin Burning Concave, and Some of the most considerable Effects of it', *PT*, 4 (1669), 986–7.

Anon, 'An Account of a Not Ordinary Burning Concave, Lately Made at Lyons', *PT*, 1 (1665), 95–8.

Anon, 'An Attempt to destroy the British Fleet' (London, 1803/4), BM 1985,0119.414.

Anon, *A Description of the Great Burning-Glass Made by Mr. Villette and his Two Sons, Born at Lyons. With some Remarks upon the surprising and wonderful Effects thereof* (London, 1718).

Anon, 'An Egyptian Solar Power Plant: Putting the Sun to Work', *Scientific American*, 107: 4 (25 January 1913), 88.

Anon, 'More Work for the Sun', *All the Year Round*, 16: 401 (5 August 1876), 490–3.

Anon, 'The Place Where Light Dwelleth', *The British Quarterly Review*, 51 (April 1870), 409–41.

Anon, *The Priestley Memorial at Birmingham, August 1874* (London: Longman, 1875).

Anon, 'A Relation of the Great Effects of a New Sort of Burning Speculum Lately Made in Germany', *PT*, 16: 188 (1687), 352–4.

Anon, 'Robertson, Artist in Ghosts', *Household Words*, 10: 253 (27 January 1855), 553–8.

Anon, '"Sun-Harnessers" Alleged Cheats', *New York Times*, 10 August 1911, p. 7.

Anon, 'The Utilisation of Sun-Power', *Chambers's Journal of Popular Literature, Science and Arts*, Issue 807 (14 June 1879), 736–77.

Anon, 'Vast Power of the Sun Is Tapped By Battery Using Sand Ingredient', *New York Times*, 26 April 1954, p. 1.

Appleyard, Rollo, 'Photo-Electric Cells', *The Telegraphic Journal and Electrical Review*, 28: 687 (23 January 1891), 124–6.

[Arbuthnot, John], *To the Right Honourable The Mayor and Aldermen of the City of London: The Humble Petition of the Colliers, Cooks, Cook-Maids, Blacksmiths, Jack-makers, Brasiers, and Others* (London, 1716).

Asimov, Isaac, *The Beginning and the End* (New York: Doubleday, 1977).

Asimov, Isaac, *The Caves of Steel* (London: Science Fiction Book Club, 1956).

Asimov, Isaac, *The Gods Themselves* (New York: Doubleday, 1972).

Asimov, Isaac, *I, Robot* (London: Ferndale, 1954).

Asimov, Isaac, 'The Last Question' (1956), in *Space Opera: An Anthology of Way-Back-When Futures*, ed. Brian Aldiss (London: Weidenfeld and Nicolson, 1974), pp. 309–22.

Asimov, Isaac, *The Roving Mind* (1983; Oxford: Oxford University Press, 1987).

Asimov, Isaac, 'Visit to the World's Fair of 2014', *New York Times*, 16 August 1964.

Ball, Robert, *The Story of the Sun* (London: Cassell, 1896).

Ballard, J. G., *Low-Flying Aircraft and Other Stories* (London: Jonathan Cape, 1976).

Barr, J. S., trans., *Barr's Buffon: Buffon's Natural History*, 10 vols (London, 1797).

Bell Labs, 'Bell System Solar Battery Converts Sun's Rays into Electricity!', advertisement in *Life*, 41: 15 (8 October 1956), p. 3.

Bernfeld, S. K., 'The Solar Menace', *Thrilling Wonder Stories*, 10: 1 (August 1937), 111–13.

Bevington, David and others, eds., *The Cambridge Edition of the Works of Ben Jonson*, 7 vols (Cambridge: Cambridge University Press, 2012).

Binder, Otto and Al Plastino, *Superman*, 146 (July 1961).

Bowers, Brian and Lenore Symons, eds., *Curiosity Perfectly Satisfyed: Faraday's Travels in Europe 1813–1815* (London: Peter Peregrinus, in association with the Science Museum, 1991).

Bradbury, Ray, *The Golden Apples of the Sun* (New York: Doubleday, 1953).

Brewster, David, ed., *The Edinburgh Encyclopaedia*, 18 vols (Edinburgh: for William Blackwood and others, 1830).

Brontë, Charlotte, *Jane Eyre*, eds. Jane Jack and Margaret Smith (Oxford: Clarendon Press, 1969).

Brutus, 'The Destruction of France by Mirrors', *The Belfast Monthly Magazine* 6: 34 (31 May 1811), 373–4.

Campbell, Thomas, *The Poetical Works of Thomas Campbell*, 2 vols (London: Edward Moxon, 1837).

Chapin, D. M., C. S. Fuller and G. L. Pearson, 'The Bell Solar Battery', *Bell Laboratories Record*, 33 (July 1955), 241–6.

Clarke, Arthur C., *The Challenge of the Spaceship: Previews of Tomorrow's World* (London: Frederick Muller, 1960).

Clarke, Arthur C., *Profiles of the Future* (1962; London: Pan Books, 1964).

Clarke, Arthur C., 'The Stroke of the Sun', *Galaxy Magazine*, 16: 5 (September 1958), 71–7.

Coburn, Kathleen and Anthony John Harding, eds., *The Notebooks of Samuel Taylor Coleridge*, 5 vols in 10 (London: Routledge & Kegan Paul, 1959–2002).

Coleridge, Samuel Taylor, *The Collected Works of Samuel Taylor Coleridge*,
    general ed. Kathleen Coburn, 16 vols (Princeton, NJ: Princeton University Press,
    1969–2002).

Cowper, William, *The Poems of William Cowper*, eds. John D. Baird and Charles
    Ryskamp, 3 vols (Oxford: Clarendon Press, 1980–95).

Davy, Humphry, *The Collected Works of Sir Humphry Davy, Bart.*, ed. John Davy, 9
    vols (London: Smith, Elder and Co, 1839–40).

Davy, Humphry, *Elements of Chemical Philosophy* (London: for J. Johnson, 1812).

Davy, Humphry, 'An Essay on Heat, Light, and the Combinations of Light', in
    Contributions to Physical and Medical Knowledge, ed. Thomas Beddoes (Bristol,
    1799), pp. 5–147.

Davy, Humphry, 'Some Experiments on the Combustion of the Diamond and Other
    Carbonaceous Substances', *PT*, 104 (1814), 557–70.

de Balzac, Honoré, *The Alkahest: Or, The House of Claës*, ed. Katharine Prescott
    Wormeley (Boston: Roberts Brothers, 1890).

de Bergerac, Cyrano, *The Comical History of the States and Empires of the Moon and
    Sun*, trans. Archibald Lovell (London, 1687).

della Porta, Giambattista, *Natural Magick*, trans. anon (London, 1658).

Drummond, William Hamilton, *The Giant's Causeway, A Poem* (Belfast, 1811).

Elliason, Olafur, 'Little Sun' (2012), <https://littlesun.com/about>.

Ericsson, John, *Contributions to the Centennial Exhibition* (New York: The Nation
    Press, 1876).

Ericsson, John, 'Solar Heat', *Nature*, 122: 2 (29 February 1872), 344–7.

Freud, Sigmund, *The Interpretation of Dreams*, trans. James Strachey, ed. Angela
    Richards (1953; London: Penguin, 1991).

Frost, Robert, *Collected Poems, Prose, & Plays*, eds. Richard Poirier and Mark
    Richardson (New York: Library of America, 1995).

Geoffroy, Etienne-François, 'Experiments upon Metals, made with the Burning-Glass
    of the Duke of Orleans', *PT*, 26 (1709), 374–86.

Gernsback, Hugo, 'Cold Light', *The Electrical Experimenter*, 6: 7 (November
    1918), 443.

Gernsback, Hugo, 'Cold Light', *Science and Invention*, 8: 9 (January 1921), 943.

Gernsback, Hugo, *Ralph 124C 41+: A Romance of the Year 2660* (Lincoln: University
    of Nebraska Press, 2000).

Gernsback, Hugo, 'The Utilization of the Sun's Energy', *The Electrical Experimenter*, 3:
    11 (March 1916), 605–6, 662–3.

Gerrard, John, 'Solar Reserve (Tonopah, Nevada) 2014', <http://www.johngerrard.n
    et/solar-reserve.html>.

Glaser, Peter E., 'Power from the Sun: Its Future', *Science*, 162: 3856 (22 November
    1968), 857–61.

Golding, William, *Lord of the Flies* (1954; London: Faber, 1962).

Hall, John, *Poems* (Cambridge, 1646).

Hamilton, Edmond, 'Space Mirror', *Thrilling Wonder Stories: The Magazine of Prophetic Fiction*, 10: 1 (August 1937), 43–51.

Hamilton, Guy (dir.), *The Man with the Golden Gun* (Eon Productions, 1974).

Harris, John, and John Theophilis Desaguliers, 'An Account of Some Experiments tried with Mons. Villette's Burning Concave, in June 1718', *PT*, 30 (1719), 976–7.

Harris, John, *Lexicon Technicum: or, an Universal English Dictionary of Arts and Sciences*, 2nd edn, 2 vols (London, 1708).

Harrison, Harry, *Skyfall* (1976; London: Panther Books, 1985).

Haydon, Benjamin Robert, *The Diary of Benjamin Robert Haydon*, 5 vols, ed. Willard Bissell Pope (Cambridge, MA: Harvard University Press, 1960–3).

Heinlein, Robert A., 'Let there be light', *Super Science Stories* (May 1940), 34–45.

Heinlein, Robert A., *The Man who Sold the Moon* (1953; London: Pan Books, 1955).

Herschel, William, 'Experiments on the Refrangibility of the invisible Rays of the Sun', *PT*, 90 (1800), 284–92.

Herschel, William, 'Experiments on the Solar, and on the Terrestrial Rays That Occasion Heat; With a Comparative View of the Laws to Which Light and Heat, or Rather the Rays Which Occasion Them, Are Subject, in Order to Determine Whether They are the Same, or Different. Part I', *PT*, 90 (1800), 293–326.

Herschel, William, 'Investigation of the Powers of the prismatic Colours to heat and illuminate Objects; with Remarks, that prove the different Refrangibility of radiant Heat', *PT*, 90 (1800), 255–83.

Herzog, Arthur, *Heat* (1977; London: Pan Books, 1978).

Heysinger, I. W., *The Source and Mode of Solar Energy Throughout the Universe* (Philadelphia: J. B. Lippincott, 1895).

Horneck, Anthony, *The Great Law of Consideration: or A Discourse*, 2nd edn (London, 1678).

Jack, Ian, and others, eds., *The Poetical Works of Robert Browning*, 15 vols projected (Oxford: Clarendon Press, 1983-date).

Jevons, William Stanley, *The Coal Question: An Inquiry Concerning the Progress of the Nation, and the Probable Exhaustion of Our Coal-mines*, 2nd edn, revised (London: Macmillan, 1866).

Josten, C. H., trans., 'A translation of John Dee's *Monas Hieroglyphica* (Antwerp, 1564)', *Ambix*, 12: 2 & 3 (June and October 1964), 84–221.

Kaempffert, Waldemar, 'Harnessing Nature: Can the Free Energy of Space be Utilized?' *Scientific American*, 108: 14 (5 April 1913), 308–9.

Kircher, Athanasius, *Ars Magna Lucis et Umbræ* (Rome, 1646).

Kircher, Athanasius, *Ars Magna Lucis et Umbræ*, 2nd edn (Amsterdam, 1671).

Lavoisier, Antoine-Laurent de, *Oeuvres de Lavoisier*, ed. Édouard Grimaux, 6 vols (Paris: Imprimerie Imperiale, 1862–93).

Le Févre, Nicaise, *A Discourse upon Sr Walter Rawleigh's Great Cordial*, trans. Peter Belon (London, 1664).

Leinster, Murray, 'The Power Planet', *Amazing Stories: Scientific Fiction*, 6: 3 (June 1931), 198–217, 227.

Lockyer, J. Norman, *The Dawn of Astronomy: A Study of the Temple-worship and Mythology of the Ancient Egyptians* (London: Cassell and Company, 1894).

Macartney, George, *An Embassy to China; being the journal kept by Lord Macartney during his embassy to the Emperor Ch'ien-lung, 1793–1794*, ed. J. L. Cranmer-Byng (London: Longman, 1962).

Mahon, Derek, *Life on Earth* (Oldcastle, County Meath, Ireland: The Gallery Press, 2008).

Mathisen, David A., '2079: A Century of Technical and Socio-Political Evolution', *Impact of Science on Society*, 29: 1 (January–March 1979), 83–91.

McEwan, Ian, *Solar* (2010; London: Vintage, 2011).

McGann, Jerome J., ed., *Lord Byron: The Complete Poetical Works*, 7 vols (Oxford: Clarendon Press, 1980–93).

Meadows, A. J., ed., *Early Solar Physics* (Oxford: Pergamon Press, 1970).

Millar, James, ed., *Encyclopaedia Britannica; Or, A Dictionary of Arts and Sciences*, 4th edn, 20 vols (Edinburgh: Andrew Bell, 1810).

Morgan, Charles, *The Burning Glass*, 2nd edn (London: Macmillan, 1955, repr. 1961).

Müller, F. Max, *Lectures on the Origin and Growth of Religion* (London: Longmans, Green and Co., 1878).

Needham, Turberville, 'Of a New Mirror, Which Burns at 66 Feet Distance, Invented by M. De Buffon', *PT*, 44 (1747), 493–5.

Nicolini, Marquis, 'Concerning the Same Mirror Burning at 150 Feet Distance', *PT*, 44 (1747), 495–6.

Packard, Winthrop, 'Power From Sunlight', *The Technical World Magazine*, 11: 4 (June 1909), 356–60.

Poliakoff, Stephen, *Blinded by the Sun & Sweet Panic* (Portsmouth: Methuen, 1996).

Pope, Charles Henry, *Solar Heat: Its Practical Applications* (Boston, MA: by the author, 1903).

Porritt, Jonathan, *The World We Made: Alex McKay's Story From 2050* (London: Phaidon Press, 2013).

Priestley, Joseph, *Experiments and Observations on Different Kinds of Air, and Other Branches of Natural Philosophy, Connected with the Subject*, 3 vols (Birmingham, 1790).

Priestley, Joseph, *The History and Present State of Discoveries Relating to Vision* (London, 1772).

Priestley, Joseph, *Scientific Correspondence of Joseph Priestley*, ed. Henry Carrington Bolton (New York: privately printed, 1892).

Rees, Abraham, ed., *The Cyclopaedia, or, Universal Dictionary of Arts, Sciences and Literature*, 45 vols (London: Longman, 1819–20).

Robertson, Étienne-Gaspard, *Mémoires Récréatifs Scientifiques et Anecdotiques*, 2 vols (Paris: by the author and Librairie de Wurtz, 1831).

Rush, Norman, *Mating* (London: Vintage, 1991).

Sedgwick, Marcus and others, *Dark Satanic Mills* (London: Walker Books, 2013).

Shelley, Percy Bysshe, *The Poems of Shelley*, eds. Michael Rossington, Jack Donovan and Kelvin Everest, 5 vols (London: Routledge, 1989–2014).

Shumaker, Wayne, ed. and trans., *John Dee on Astronomy. Propaedeumata Aphoristica (1558 & 1568)* (Berkeley: University of California Press, 1978).

Shuman, Frank, 'Feasibility of Utilizing Power from the Sun', *Scientific American*, 110: 9 (28 February 1914), 179.

Shuman, Frank, 'Power from Sunshine: A Pioneer Solar Power Plant', *Scientific American*, 105: 14 (30 September 1911), 291–2.

Shuman, Frank, 'Sun-power Plants Not Visionary', *Scientific American*, 110: 26 (27 June 1914), 519.

Siemens, Charles William, 'On the Conservation of Solar Energy', *Proceedings of the Royal Society of London*, 33 (1882), 389–98.

Smith, Willoughby, 'The Action of Light on Selenium', *Journal of the Society of Telegraph Engineers*, 2 (1873), 31–3.

Stapledon, W. Olaf, *Last and First Men: A Story of the Near and Far Future* (London: Methuen, 1930).

Stapledon, W. Olaf, *Star Maker* (London: Methuen, 1937).

Stewart, Balfour and J. Norman Lockyer, 'The Sun as a Type of the Material Universe', Parts I and II, *Macmillan's Magazine*, 18 (July 1868), 246–57, 319–27.

Swift, Jonathan, *Gulliver's Travels*, ed. David Womersley (Cambridge: Cambridge University Press, 2012).

Tamahori, Lee (dir.), *Die Another Day* (MGM, 2002).

Thomson, William, 'On the Age of the Sun's Heat', *Macmillan's Magazine*, 5 (March 1862), 388–93.

Thomson, William, 'On the Mechanical action of Radiant Heat or Light; On the Power of Animated Creatures over Matter; On the Sources available to Man for the production of Mechanical Effect', *Proceedings of the Royal Society of Edinburgh*, 3 (1857), 108–13.

Thomson, William, 'On the Mechanical Energies of the Solar System' (1854), *Transactions of the Royal Society of Edinburgh*, 21 (1857), 63–81.

Tyndall, John, *Essays on the Use and Limit of the Imagination in Science* (London: Longmans, Green, and Co, 1870).

Tyndall, John, *Heat Considered as a Mode of Motion* (London: Longman, 1863).

Tyndall, John, 'On Force', *Philosophical Magazine*, 24 (1862), 57–66.

Wagner, Phoebe and Brontë Christopher Wieland, eds., *Sunvault: Stories of Solarpunk and Eco-Speculation* (Nashville, TN: Upper Rubber Boot, 2017).

Wells, H. G., 'An Excursion to the Sun', *Pall Mall Gazette*, 58 (6 January 1894), 4.

Wells, H. G., 'Scientific Research as a Parlour Game', *The Saturday Review*, 79 (20 April 1895), 516.

Wells, H. G., 'The Sun God and the Holy Stars', *Pall Mall Gazette*, 58 (24 February 1894), 3.

Wells, H. G., *The Time Machine*, ed. Roger Luckhurst (Oxford: Oxford University Press, 2017).

Wells, H. G., *The War of the Worlds*, eds. David Y. Hughes and Harry M. Geduld (Bloomington: Indiana University Press, 1993).

Wells, H. G., *The World Set Free: A Study of Mankind* (New York: E. P. Dutton, 1914).

Woolf, Virginia, *Moments of Being*, ed. Jeanne Schulkind, rev. Hermione Lee (London: Pimlico, 2002).

Young, Charles Augustus, *The Sun* (New York: D. Appleton, 1881).

Zola, Émile, *Labor: A Novel*, trans. anon (New York: Harper, 1901).

Secondary sources

Ashworth, William B., 'Light of Reason, Light of Nature. Catholic and Protestant Metaphors of Scientific Knowledge', *Science in Context*, 3: 1 (March 1989), 89–107.

Barnham, Keith, *The Burning Answer: A User's Guide to the Solar Revolution* (London: Weidenfeld & Nicolson, 2014).

Barrett, Ross and Daniel Worden, eds., *Oil Culture* (Minneapolis, MN: University of Minnesota Press, 2014).

Bartels, Daniel, 'George Cove's Solar Energy Device', *Material Culture Review*, 46 (Autumn 1997), 45–50.

Bate, Jonathan, *The Song of the Earth* (London: Picador, 2000).

Beer, Gillian, '"The Death of the Sun": Victorian Solar Physics and Solar Myth', in *The Sun Is God: Painting, Literature and Mythology in the Nineteenth Century*, ed. J. B. Bullen (Oxford: Clarendon Press, 1989), pp. 159–80.

Beer, Gillian, *Open Fields: Science in Cultural Encounter* (Oxford: Oxford University Press, 1996).

Behrman, Daniel, *Solar Energy: The Awakening Science* (London: Routledge & Kegan Paul, 1979).

Cantor, G. N., 'Weighing Light: The Role of Metaphor in Eighteenth-Century Optical Discourse', in *The Figural and the Literal: Problems of Language in the History of Science and Philosophy, 1630–1800*, eds. Andrew E. Benjamin, G. N. Cantor and John R. R. Christie (Manchester: Manchester University Press, 1987), pp. 124–46.

DeLoughrey, Elizabeth, 'Heliotropes: Solar Ecologies and Pacific Radiations', in *Postcolonial Ecologies: Literatures of the Environment*, eds. DeLoughrey and George B. Handley (Oxford: Oxford University Press, 2011), pp. 235–53.

Dickson, David, *Alternative Technology and the Politics of Technical Change* (London: Fontana, 1974).

Dunaway, Finis, *Seeing Green: The Use and Abuse of American Environmental Images* (Chicago: The University of Chicago Press, 2015).

Garrard, Greg, 'Solar: Apocalypse Not', in *Ian McEwan: Contemporary critical perspectives*, ed. Sebastian Groes, second edn (London: Bloomsbury, 2013), pp. 123–36.

Gold, Barri J., *Thermopoetics: Energy in Victorian Literature and Science* (Cambridge, MA: MIT Press, 2010).

Goldberg, Benjamin, *The Mirror and Man* (Charlottesville, VA: University Press of Virginia, 1985).

Goodall, Chris, *The Switch* (London: Profile Books, 2016).

Grabes, Herbert, *The Mutable Glass: Mirror-Imagery in Titles and Texts of the Middle Ages and English Renaissance* (Cambridge: Cambridge University Press, 1982).

James, Frank A. J. L., 'Thermodynamics and Sources of Solar Heat, 1846–1862', *British Journal for the History of Science*, 15: 2 (July 1982), 155–81.

Jenkins, Alice, 'Humphry Davy: Poetry, Science and the Love of Light', in *1798: The Year of 'Lyrical Ballads'*, ed. Richard Cronin (Houndmills: Macmillan, 1998), pp. 133–50.

Kidwell, Peggy Aldrich, 'Prelude to Solar Energy: Pouillet, Herschel, Forbes and the Solar Constant', *Annals of Science*, 38 (1981), 457–76.

Klein, Naomi, *This Changes Everything: Capitalism vs. the Climate* (London: Allen Lane, 2014).

Kragh, Helge, 'The Source of Solar Energy, ca. 1840–1910: From Meteoric Hypothesis to Radioactive Speculations', *The European Physical Journal*, 41 (2016), 365–94.

Kryza, Frank T., *The Power of Light: The Epic Story of Man's Quest to Harness the Sun* (New York: McGraw-Hill, 2003).

Lee, Hermione, 'A Burning Glass: Reflection in Virginia Woolf', in *Virginia Woolf: A Centenary Perspective*, ed. Eric Warner (Basingstoke: Macmillan, 1984), pp. 12–27.

Leggett, Jeremy, *Half Gone: Oil, Gas, Hot Air and the Global Energy Crisis* (2005; London: Portobello Books, 2006).

Macdonald, Graeme, 'Improbability Drives: The Energy of SF', *Paradoxa*, 26 (2014), accessed via <http://strangehorizons.com/non-fiction/articles/improbability-drives-the-energy-of-sf/>.

MacDuffie, Allen, *Victorian Literature, Energy and the Ecological Imagination* (Cambridge: Cambridge University Press, 2014).

MacDuffie, Allen, 'Victorian Thermodynamics and the Novel: Problems and Prospects', *Literature Compass*, 8: 4 (2011), 206–13.

Madrigal, Alexis, *Powering the Dream: The History and Promise of Green Technology* (Cambridge, MA: Da Capo Press, 2011).

Malm, Andreas, *Fossil Capital: The Rise of Steam Power and the Roots of Global Warming* (London: Verso, 2016).

Melchior-Bonnet, Sabine, *The Mirror: A History*, trans. Katharine H. Jewett (New York: Routledge, 2001).

Middleton, W. E. Knowles, 'Archimedes, Kircher, Buffon, and the Burning-Mirrors', *Isis*, 52: 4 (December 1961), 533–43.

Monbiot, George, *Heat* (2006; London: Penguin, 2007).

Morton, Timothy, 'Shelley's Green Desert', *Studies in Romanticism*, 35: 3 (1996), 409–30.

Nye, David E., *Consuming Power: A Social History of American Energies* (Cambridge, MA: The MIT Press, 1998).

Nye, David E., 'Energy Narratives', *American Studies in Scandinavia*, 25 (1993), 73–91.

Patterson, Bill, 'A Study of "Let There Be Light"', *Heinlein Journal*, 4 (January 1999), 17–22.

Perlin, John, *From Space to Earth: The Story of Solar Electricity* (Ann Arbor, MI: Aatec Publications, 1999).

Nye, David E., *Let It Shine: The 6,000-Year Story of Solar Energy* (Novato, CA: New World Library, 2013).

Plassmeyer, Peter and others, *Ehrenfried Walther von Tschirnhaus (1651–1708): Experiment emit dem Sonnenfeuer* (Dresden: Staatliche Kunstammlungen Dresden, 2001).

Schivelbusch, Wolfgang, *Disenchanted Night: The Industrialisation of Light in the Nineteenth Century* (Oxford: Berg, 1988).

Simms, D. L. and P. L. Hinkley, 'Brighter than How Many Suns? Sir Isaac Newton's Burning Mirror', *Notes and Records of the Royal Society of London*, 43 (1989), 31–51.

Smeaton, W. A., 'Some Large Burning Lenses and Their Use by Eighteenth-Century French and British Chemists', *Annals of Science*, 44 (1987), 265–76.

Solnick, Sam, *Poetry and the Anthropocene: Ecology, Biology and Technology in Contemporary British and Irish Poetry* (Abingdon, Oxon: Routledge, 2016).

Szeman, Imre and Dominic Boyer, eds., *Energy Humanities: An Anthology* (Baltimore: John Hopkins University Press, 2017).

Szulakowska, Urszula, *The Alchemy of Light: Geometry and Optics in Late Renaissance Alchemical Illustration* (Leiden: Brill, 2000).

Tattersdill, Will, *Science, Fiction, and the Fin-de-Siècle Periodical Press* (Cambridge: Cambridge University Press, 2016).

Underwood, Ted, 'How did the Conservation of Energy become "the highest law in all science?"', in *Repositioning Victorian Sciences: Shifting Centres in Nineteenth-Century Thinking*, ed. David Clifford and others (London: Anthem Press, 2006), pp. 119–30.

Underwood, Ted, *The Work of the Sun: Literature, Science, and Economy, 1760–1860* (New York: Palgrave Macmillan, 2005).

Warner, Marina, *Phantasmagoria: Spirit Visions, Metaphors, and Media into the Twenty-first Century* (Oxford: Oxford University Press, 2006).

Westfahl, Gary, *Islands in the Sky: The Space Station Theme in Science Fiction Literature*, 2nd edn (1996; San Bernardino, CA: Borgo Press, 2009).

Worth, Aaron, 'Imperial Transmissions: H. G. Wells, 1897–1901', *Victorian Studies*, 52 (2010), 65–89.

Yaeger, Patricia, 'Literature in the Ages of Wood, Tallow, Coal, Whale Oil, Gasoline, Atomic Power, and Other Energy Sources', *PMLA*, 126: 2 (2011), 305–26.

Zajonc, Arthur, *Catching the Light: The Entwined History of Light and Mind* (Oxford: Oxford University Press, 1993).

Index

Lightning Source UK Ltd.
Milton Keynes UK
UKHW022206040320
359783UK00003B/119